Farmlife

From Farm to Table and New Country Culture

gestalten

Introduction

Food for Thought

There is a global movement to encourage greater engagement with our food systems and an understanding that to do so is to protect the environment, public health, local communities, and animal welfare. Learning how to live a more sustainable lifestyle and make healthier, more conscious food choices offers us greater harmony with nature in hopes of reducing stress on ourselves and on the planet.

We collaborated with Food Studio in Oslo, Norway, to bring you the stories of passionate people—grandmas and young entrepreneurs alike—who are working to preserve traditional practices and recipes, harvest their own food, and get it from the farm to our tables: the people empowering us to make better choices in our day-to-day lives whether shopping, cooking, or sharing a meal.

City dwellers need not fear! Regardless of your proximity to a farm (or your desire to visit one), there are simple yet meaningful ways to join this global effort. Read on as Cecilie Dawes, founder of Food Studio, shares her personal journey toward becoming a food ambassador.

"We don't know where our food is coming from, what is in it, and how the ingredients will affect both our own health and the health of the planet."

Life has taught me that the most beautiful gifts are the ones you open with your heart even when you don't yet know the outcome. As humans, we have a natural inclination to believe that we control our path. From childhood, we are taught that a good education will lead to a good job, which will give us a secure income to provide for and feed our household. But the complexities of life are far too great for a teenager to fully grasp the consequences of early educational choices.

To me it seems that the challenge of building an authentic life—a uniquely personal thing for each of us—is actually based more on gut feeling and a sense of received wisdom. While reasoning and intellect are foundational to decision making and the human experience, they are only part of the story, and sometimes we have to let these impulses go to see the whole picture.

Some parts of the journey of my life have felt quite chaotic, accidental, and aimless, but my gut has always known some things are right for me, regardless. The journey of creating Food Studio has always been one of those things. In the beginning, I was clueless as to what it was going to be. Initially it was a departure ticket out of a food system that I was a part of but had come to perceive as fundamentally wrong. I thought I was starting a business, but it became so much more; it became the key to being myself fully and living the life I believe in, together with people that share the same philosophy of life, those who are longing for a community without the limitations existing in today's society. Food Studio has become a collective of people who rally around the desire to live a resource-friendly life: researching, experimenting, living, experiencing, and developing new philosophies of sustainability.

> "What I see in almost every farmer is a true passion and respect for the circle of life."

As for most people, my journey into the world of food started in the family. Our family led busy working-class lives. When it was my turn to make dinner at the age of ten, I found the cabinets filled with the instant sauce mixes and powders of the time, the latest solution for a generation of full-time working women. During my teenage years, I fed my body processed white bread, sweets, chocolate, and soda, throwing away my beautiful homemade lunches, only to end up with fatigue and headaches at the end of the day. I was completely removed from the food cycle.

There are still so many teenagers like the one I was back then. We don't know where our food is coming from, what is in it, and how the ingredients will affect both our own health and the health of the planet. We have lost the connection to our food, and the rituals that come with preparing it.

My fondest childhood memories are from days spent with my grandmother, blanching cabbage leaves and wrapping them around minced meat to make my favorite dish at the time—cabbage rolls. I can also remember the sensation of being woken up in the middle of the night to my uncle's freshly caught brown trout prepared simply at our family cabin in the mountains.

I studied economics and resource management at the Agricultural University of Norway, and, through courses in food culture as well as a master's in food science and rural development, I gained insight into the "grocerization" of Norway. It was when I first began working for a chain convenience store and became part of the central national procurement system that I fully realized how difficult it was for small producers to enter the market with sustainable products. It was painful to be part of a system that denied shelf space to food manufacturing entrepreneurs while simultaneously stealing ideas from their product pitches. Price and volume took precedent over quality, nutrition, and health benefits. At the age of 15, as I surrounded myself with the comforting yet unsettling aroma of hash browns and chicken nuggets on my first day of work at the local fast food franchise, I had no idea that my education and professional career would circle ever closer into the realms of food and communication.

As my own instinctual food philosophy pulled me more and more in the opposite direction, it became exhausting to work within the limitations of the current system. Lengthy discussions on short-term profits, deals with larger suppliers, and debates on whether or not to reintroduce hot dogs drained me. At the time, my boss and I made a good team, but after having to assume some of his responsibilities while he was on extended sick leave, I knew I was nearing my time's end.

I booked a three-week-long trip to Buenos Aires to clear my head. When I returned, I spent half a year pulling together my network, meeting new and inspiring people, and thinking deeply about the meaning of sustainability. I ended up with a business plan for Food Studio and applied for support from Innovation Norway. Soon, Food Studio was my full-time job. The seeds I planted while sitting and sketching in Buenos Aires began to blossom and take on a life of their own.

Food Studio

From the very beginning, a team of photographers, writers, designers, artists, and interdisciplinary academics congregated around the project. In striving to find the few, simple words to best describe our vision, the term "Food Empathy" was kicked around and started to feel more and more like an all-encompassing description for Food Studio.

By "Food Empathy", we meant an essential understanding of what food is, where it comes from, the journey it had undertaken, the resources harnessed from sun, soil, and grass, and how it enriched not just our souls, but also our bodies. It is what nourishes us and what we are made of, and it returns to nature when we throw it away

Below—The greenhouse was built by students. As is the case for all elements of a farm, it needs maintenance, and the Food Studio team has plans to restore it.

or digest it. In short: an understanding of the great wheel of sustenance and how it comes full circle.

Food Studio believes that "Food Empathy" is developed through a combination of formative sensory experiences and through the natural process of our own maturity in which, ideally, the self gives way to a broader awareness of our common cultural heritage. Food plays a huge part in this. And experiential learning is essential for development and knowledge to move from theory to practice. When something is experienced and perceived with all the senses, connections that otherwise might be lost are instead created.

Gaining tangible insight into how things fit together also inspires change. For example, following a tomato plant from seeds to sweet, ripe tomatoes, or converting an old heel of bread into bread pudding gives us a deeper understanding of and connection to our food. And enjoying a weekend meal of meat from a local farmer brings us closer to a respect for our food and the environment.

At Food Studio, we create experiences for both mind and body by pairing passionate experts in their respective fields, including farmers, chefs, scientists, activists, and hobbyists with people from all walks of life. We bring communities together for lectures, discussions, harvesting, foraging, and wandering the landscape.

Above—Food Studio founder Cecilie Dawes in Hegli Farm's wild garden. **Below**—The autumn harvest of kale, zucchini, chives, and mangold.

Finally, we create and enjoy beautiful meals together, taking in and digesting what we have learned and hopefully integrated. At the same time, our network of journalists and researchers document our gatherings with words, films, and photos to share our initiatives with the wider world. Our team is ever-changing, location-independent, and nomadic, our work is project-based.

Since the beginning, we have worked with a wide variety of farmers both in our own region and in the areas we visit around the world. It is fascinating to see how much they share in terms of the struggles against the system of conventional farming and food quality authorities, and especially against the imbalance of powers. At the same time, a new generation of farmers is rapidly creating a movement and building their own networks for collaboration with parallel distribution channels alongside the existing structures.

Alternative platforms such as community-supported agriculture (CSA) revitalize farms as a group effort, creating value not just in farm produce but also in working and learning as a collective. They create new products and distribution systems in the retail spectrum, (home deliveries, for example), and have an outstanding dedication to and understanding of quality to meet the demands from a growing number of professional chefs seeking farm collaborations. In this way, farmers are able to build bridges from themselves to the city. What I see in almost every farmer is a true passion and respect for the circle of life. The more we learn about the complexities involved

in making a living as a farmer, the more respect we gain for those who have found the key to doing this in a balanced way. When people consider the price of food, the gratitude of "Food Empathy" should be considered; the deeper we continue to educate ourselves about what it really takes to bring a meal to fruition, the more we will bow with respect.

As in other industries, the most successful farmers are often the ones who understand social networks and utilize the power of narrative to bring their produce to markets. The irony is that farm life itself is a life spent mostly apart from smartphones or computers, acted out by physically working in the fields, barns, hen houses, and stables.

Storytelling has moved from around campfires or in community councils to online forums. The horse and cart have been replaced by Facebook and Instagram. Most farmers don't have the time or resources to dedicate to these platforms to keep their stories alive and public. Naturally, social networking skills and storytelling don't usually fall into the standard skillset in agriculture. This is why a book such as this one is so important, introducing a broader audience to and advocating for the passionate farmers who spend the days with their hands in the soil. In many ways, the food industry functions as an oligarchy. There are a few major players, and they control or deeply influence all elements of the food path from nature to the counter at the supermarket. That said, the industrialization of food has been an amazing innovation that has helped to reduce costs, enhance productivity, and exchange food across a variety of cultures.

The benefits are undeniable. But these innovations have been brought about at a social cost. Every day, I learn more about the potential dangers of commercial fertilizers, pesticides, monocultures, flavor enhancers, preservatives, bulking agents, packaging, waste, and long transport hauls—the list goes on and on. As consumers, we have become alienated from our natural inclination to have a relationship with what we eat, and we are exposed to a whole range of unknowns in the food that we buy. Many people in urban communities today have never even met a farmer.

From this situation comes a reactionary parallel market system created by farmers, producers, and innovators, where following the product all the way to market brings greater value to them. Many farmers only sell to customers they have a relationship with, and platforms are being created to connect farmers and chefs so they can enter into cooperations that are nonexistent in conventional systems. Farmers' older market models have evolved into more creative spaces where young farmers not only sell their vegetables in bulk, but also ferment, pickle, cure, and experiment.

Opposite—Hegli Farm is owned by Sidsel Sandberg (pictured left), who has lived and worked here for over 50 years. **Left**— Food Studio founder Cecilie Dawes. **Right**—Julius Maske, a gardener at Hegli Farm and a student in the BING program, an educational initiative for a new generation of farmers.

This innovative platform with instant feedback from customers is a highly effective way to develop new products. In the following pages, you will be introduced to a variety of these models, from chef–farmer collaborations, to innovative foragers, beekeepers, and florists, or farms used as a guesthouse, an organic retreat, or a platform for an apprenticeship.

The disconnect between consumer and systems of food production led me to embark on a one-year research-based experiment. And for one year I "practiced what I preached." There are really good frameworks out there such as the UN Convention on Climate Change Collaborations; with the help of scientists and researchers, I made a personal plan, a set of guidelines to follow and live by over time. The first step was to stop flying, and get rid of my car. I stopped purchasing consumer goods and stopped shopping for food in normal grocery stores; instead I strived to be as self-sustained as possible, and I used Oslo Cooperative, a collective that trades directly with farms, as my main source of vegetables in the low growing season. I volunteered for two urban gardening projects, where I traded work for vegetables, and I transformed as much of my city balcony as possible into an edible garden. I also began to use a Bokashi composting system at home.

The fat sources on my daily menu were butter, rapeseed oil, and leftover animal fat. Local honey was my sweetener, and I used meats to spice up vegetable dishes. I started a routine of boiling salt from seawater, making apple cider vinegar, fermenting vegetables, and making sourdough bread. I considered alternative sustainable housing options such as a yurt or a wood cottage. I was well aware that I was walking on the fringe, but I also knew that my somewhat utopian starting point would be moderated by practice and the constraints of my social and professional lives.

Finding Hegli Farm

I learned of Hegli Farm in the early phases of Food Studio through a mutual friend, teacher, and great inspiration Linda Jolly. Sidsel Sandberg is the farmer at Hegli, and she has owned and worked the farm since 1966. In following the journey of these two women and learning from their experiences, I realized that the way I had designed Food Studio's teaching strategies and gatherings is similar to how Linda and Sidsel work with their students: emphasizing beautiful physical surroundings, waking the senses, and creating an atmosphere of warmth, respect, and mindfulness. It's incredibly inspiring to cooperate and learn from these friends and mentors, aspiring toward their courage and determination in materializing their visions, wherever that may lead.

Sidsel is now 75 and her body is telling her that it is time to recruit some help. She can do the morning and evening work in the barn, but everything else is too much. None of her four children want to run the farm as it is today. Sidsel's situation is not unusual for an older farmer. All over the country, the same thing is happening. It would be easy to lease the land, but Sidsel wants the farm to be kept and run as a whole entity. She believes that in times like these it is sensible to have access to land, and this is her life's work: not just anybody can carry on her legacy.

Sidsel's farm was never just a regular farm with cattle, hens, and vegetables. For 19 years, Sidsel was doing educational work through a garden and a kitchen for students. Hegli Farm was part of a project called Levande Skule (Living School)—farms as an educational resource. Throughout their three years in middle school, students worked for a total of four weeks on the farm. They worked for one week at a time during different seasons, enabling them to participate in and observe the circle of life and how animals and plants grow.

The students learned about sowing, harvesting, gathering, butchering, preserving, and pickling. During the last week of the project, they made a feast for their grandparents. They were responsible for everything from invitations, menus, gathering and preparing the food, and organizing the feast itself. "In this school, they really learned about life," says Sidsel.

A genuine life from micro to macro, from earth to table. The students also learned a lot about cooperation by working together as a community, and most students have said that they really enjoyed the experience, both the theoretically inclined and those who have practical passions. On a farm, there are tasks and challenges for all

> "As consumers, we are often confronted with our footprint when it comes to transport and housing, but few of us are as aware that the choices we make each day on our plates actually create the biggest footprint of all."

kinds of people, according to Sidsel. Sidsel tells me that she has just read Naomi Klein's book *No Is Not Enough*. Klein says that there is no point in complaining about things changing in ways we don't like. She says it is time to think strategically about how to make things better. We need to focus not on what we say no to, but what we say yes to. But what are we saying yes to? What is it we want? "That is what we need to discover," says Sidsel. She and I agree that these new times demand new ways of thinking and acting. And there is never just one solution. When the problem is complex, there are many solutions. But fellowship and community are values we both believe in—doing things together; collective mindsets, collective solutions.

Farmers of the Future

Sidsel is open to the idea that her farm could be leased to a collective who is excited about exploring new ways to run a versatile, organic farm. She envisions a site with educational programs. If everyone contributes their individual skills, time, resources, knowledge, and talents, could the collective create a system where no one owns the farm, but everyone takes care of it? Because who really owns the earth? Is it not up to us to look after her as best we can and be grateful for all that she gives us?

My meetings with Linda and Sidsel, and the process that followed, sparked the beginnings of Hegli Sjølbergingsbruk, a farm with a core mission of self-sustenance and dissemination of knowledge. At an early stage, we invited the local community in the nearby Norwegian town of Nannestad, as well as the city people of Oslo and chefs we know. We asked them for input and ideas on how to reach our goals. What would be relevant for them to learn more about? What produce is not available in the existing markets? What form of community will build bridges from the farm to the city, from the rural to the urban? An interesting mix of people came together out of this meeting, old friends and curious strangers wanting to learn. One participant had a microhouse but no land. One couple had been a part of a similar project in Northern California called the Farm Stand Art Space. Their community project started with bringing produce to the city, and in doing so, people started cooking together, organizing potlucks, and drafting mealplans from the fresh produce in the boxes from the farm. The liveliness of this community, where people offered up the skills they had, from musicians playing, to chefs sharing their recipes—being open to all kinds of people—was something that deeply inspired us. The gap between consumers and the production system may be one of the biggest collective challenges we face. As consumers, we are often confronted with our footprint when it comes to transport and housing, but few of us are as aware that the choices we make each day on our plates actually create the biggest footprint of all, and there is a very polarized and confusing discourse.

On one end of the spectrum we find the traditional environmentalists, on the other the technology optimists, and in between, an ocean of opinions. Politicians are trying to toe the line between all these schools of thought. None of them hold the whole truth, but the solutions we need in order to meet the challenges of the future are to be found in open dialog and collaborations between all actors. Building bridges seems more important than ever. If you think that you are one of the farmers of the future and you want to learn more about how you can move in this direction, there are many different paths. The traditional school system offers a wide range of possibilities, even though there has been a problematic gap between conventional and closed-loop farming methodology, with the conventional direction having been at the forefront for a long time.

Above—A table of food made at the farm—apple sauce, honey, butter, fresh cheese and sourdough bread baked in the wood-fired oven.
Left—Living by the seasons in the Nordic countries forces a farmer to preserve much of the summer harvest to spice up the winter meals. Hegli Harm has no refrigerator, but a cellar where they do the canning.

But this is changing, and traditional farming and gardening schools are now incorporating the latter into their focus. Learning through practice is also a very important part of this education, so many young people learn through volunteering at farms. There are several international options available. The WWOOF (World Wide Opportunities on Organic Farms) organization and the Workaway platform are two of these. The Biodynamic Association has a program for young people called BING (Biodynamic Initiative for the New Generation), and the permaculture movement has its own educational programs and courses. Most of these programs build on an apprenticeship model with a shoulder-to-shoulder mentoring philosophy.

In this book, there is opportunity to more deeply explore the stories of the people behind your food—stories that are usually hard to access. These stories build knowledge and foster "Food Empathy". We invite you to find inspiration, big or small—which might just lead to a change of heart or practice.

—Cecilie Dawes, founder of Food Studio

Köping, Sweden

Returning Home to Tavsta Farm

Linnéa and **Pelle Holst** have returned to the land where Linnéa grew up to raise their daughter among vast fields and wild woods.

It started with a wish to live closer to nature. "We have always known that we want to live in the country." Linnéa Holst is a photographer and jeweler, and her husband Pelle is a former circus and theater performer who has since turned his hand to carpentry with a focus on traditional techniques.

Together they reside at Tavsta Farm, a sprawling 30-acre property set on the edge of freshwater Lake Mälaren in Sweden's south. Linnéa grew up at Tavsta where her parents still live, while Pelle was raised on a small farm in the west of Sweden before moving to the city. "I grew up with woodworking, gardening, hunting, and painting all around me, and always had a love for woods. To me, a forest offers comfort, rest, and a huge food supply," Pelle says.

Each time the couple visited Linnéa's parents, their walks on the property led to one particular crumbling cottage. "The house triggered our imagination and we decided to give it a chance," Pelle says. Driven by the desire to spend more time in nature, they began to slowly transform the 1832-built timber house. At first it was a weekend project, a residence they were building for summer vacations. But when their daughter was born, they decided to move to the farm permanently.

With a tiny budget and generous help from friends and family, they expanded the house to twice its size using recycled and natural materials. From their character-filled sanctuary, they look out to vast stretches of fields to the north and dense woods of oak, birch, pine, hazel, and maple trees to the south.

Tavsta Farm—Köping, Sweden

Tavsta Farm—Köping, Sweden

The couple planted a permaculture garden to grow vegetables and began to pour love into a neglected 200-year-old orchard. "The soil is clay-based and quite heavy, therefore digging is hard and at times impossible. But if taken care of properly, it's wonderful to work with." To conquer the challenges of the soil, Pelle and Linnéa grow myriad perennials, along with small beds of annuals like carrots, beans, corn, sugar snap peas, tomatoes, cucumbers, and zucchini. They find the fruit trees and berries mostly take care of themselves, providing a bountiful yield to preserve as jams or freeze for smoothies. "Eating organic and healthy food is important to us," Linnéa says. "I like to know where the food we eat comes from, where it is grown, and who has grown it. So it's nice to do a lot of the work ourselves."

While Pelle builds houses by day, Linnéa works at her store in nearby Köping, which stocks organic, sustainable, and fair-trade products. "When we get home, we make dinner from our own vegetables and spend time with our daughter. During the weekends we are outdoors as much as possible, working on the farm or in the woods," Linnéa says.

As their homesteading journey continues, they plan to expand the garden, build greenhouses, and open a bed and breakfast and small shop from which to sell their produce. "Our farm is a lifelong project," says Pelle. But they're already witnessing the beautiful effects that living in connection with the land can have on a human through their daughter's eyes. "She is three years old and already knows a lot of birds just by their song. She can distinguish some edible mushrooms and knows the names of many plants and animals," share the proud parents.

Linnéa believes everyone can take small steps toward living a more environmentally minded existence. "We want to do our part to take care of the earth and to live as resourcefully as possible. We only have one planet and we need to take a stand and be responsible for making the earth feel good again."

Seasonal bounty—Pelle and Linnéa eat with the seasons and practice traditional homesteading methods like drying and preserving fruit to ensure they have enough food throughout the dormant winter.

Sponge Cake

with Elderflower Glaze

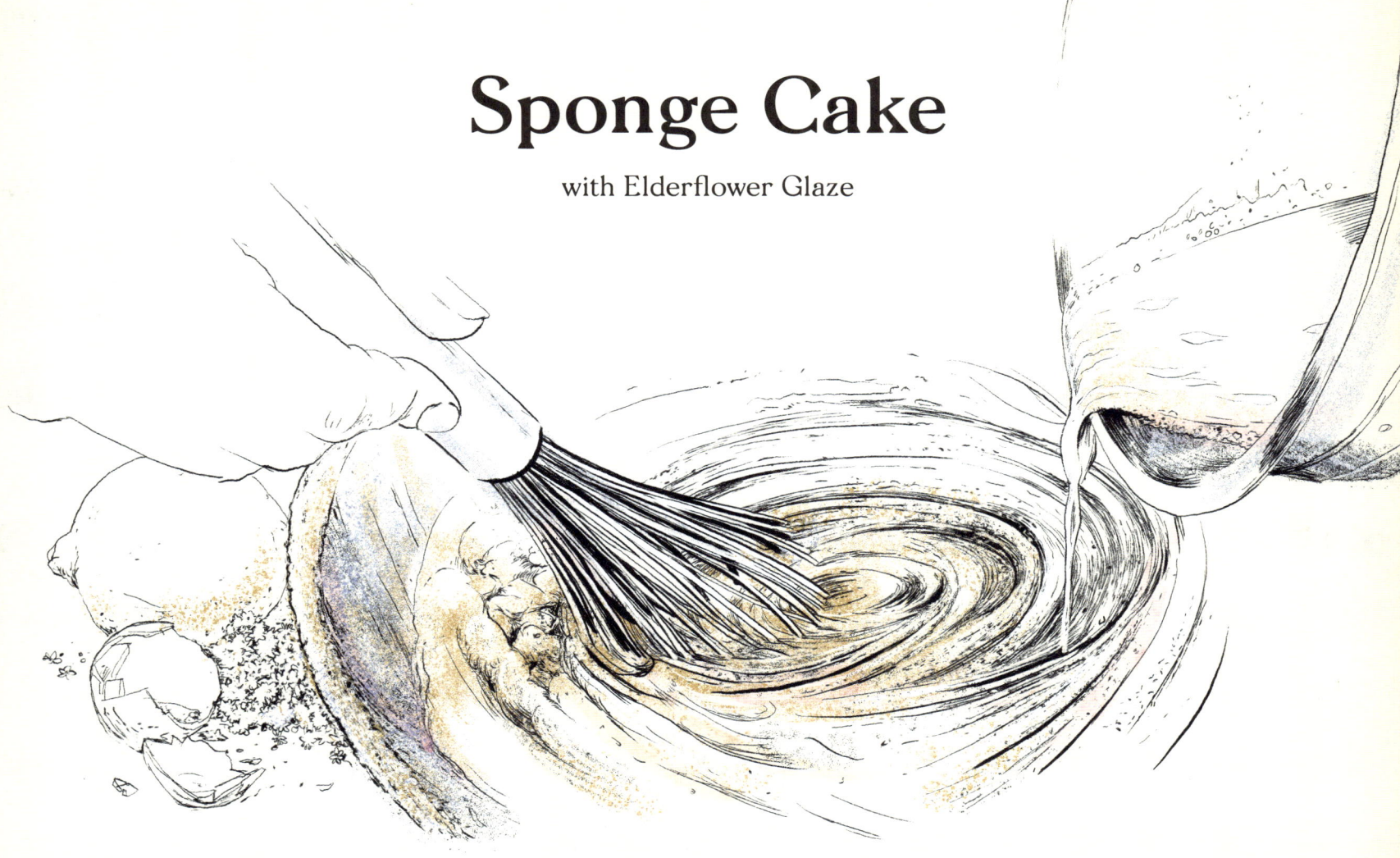

Serves 8–10

Ingredients

3 **eggs**
1 1/4 cups (240 g) granulated sugar
3 **drops** vanilla extract
Zest of 1 organic lemon
1/4 cup (50 g) butter, melted, plus extra for greasing
1/2 cup (100 ml) milk
1 1/3 cups (180 g) flour
2 **teaspoons** baking powder
Pinch of salt
3/4 cup (100 g) powdered sugar
4 **tablespoons** elderflower concentrate
Edible flowers, for garnish (optional)

Equipment

10-in (22-cm) cake tin
2 bowls
Whisk
Flour sifter or sieve

Instructions

1. Preheat oven to 180 °C (350 °F) fan-forced and grease the cake tin. In a bowl, whisk the eggs, sugar, vanilla, and lemon zest until white and very fluffy. Pour the milk into the melted butter and mix well, then fold into the fluffy egg mixture until just combined.

2. Sift together the flour, baking powder, and salt, and gently fold into the batter. Pour the batter into the prepared cake tin and bake until a toothpick comes out clean, about 30 minutes (or a little longer if your oven does not have the fan-forced option).

3. Let your cake cool in the tin on a rack before turning it out onto a plate. Do not skip this step, or the cake may fall apart.

4. In another bowl, mix the icing sugar and elderflower juice until they reach an egg-white consistency, adjusting with more sugar or juice as needed. Pour the glaze over the cooled cake and garnish with a few edible flowers if you have any in the garden: nasturtiums, cornflowers, calendula petals, elderflowers...

Enjoy!

A Tuscan Estate with Sustainable Values

– Spannocchia has become an exemplar for conservation in the region with its organic farm and educational programs

Siena, Tuscany, Italy

Spannocchia was purchased in 1925 by Florentine writer Delfino Cinelli, whose granddaughter Francesca Cinelli now runs the estate with her American husband Randall Stratton.

Like a beautiful stone fortress that overlooks the undulating green Tuscan landscape, the 1,100-acre Spannocchia estate is all about conservation. Whether it's adherence to traditional materials during renovation or their certified organic farm that raises endangered breeds of local animals such as the Cinta Senese pig, Spannocchia is a self-sustaining agent in a region beleaguered by gentrification.

Dishes are made from the ingredients grown at their sprawling permaculture garden, and wine and olive oil are produced from their vineyard and orchard. Guests enliven the pastoral estate and are invited to take part in the Friends of Spannocchia non-profit educational events, including workshops in olive oil, Italian wine, pasta-making, or slow living and tasting. For a more immersive experience, the farm internship program is a thoroughly hands-on introduction to farming, agricultural tourism, and sustainability.

Guests enliven the pastoral estate, and they can take part in the Friends of Spannochia non-profit educational events, including workshops in olive oil, Italian wine, pasta-making, or slow living and tasting.

The farm continues to experiment with new techniques such as synergistic gardening, where soil health is maintained through plant selection and recycling rather than digging or the use of fertilizers.

Spannocchia—Siena, Tuscany, Italy

A Seed-to-Table Movement

– **Tim Mountz**'s Happy Cat Farm actively preserves heirloom seeds with a focus on sustainability

Kennett Square, Pennsylvania, USA

The story of Tim Mountz and Happy Cat Farm began with an old jar of beans. While settling the estate of Mountz's grandfather, his grandmother uncovered the jar, setting their lives down a whole new path. With the help of food historian William Woys Weaver, Mountz discovered the centuries-old origins of the beans, rife with stories of Native Americans and runaway slaves. And so their mission was clear: to save these heirloom seeds, which have become almost entirely displaced by modern agriculture.

The small sustainable farm and nursery specializes in heirloom vegetable seeds, with over 200 varieties of heirloom tomatoes and produce. With a passion for teaching others how to grow their own food sustainably, Mountz's seed-to-table farming philosophy is an act of patience with a deep respect for the environment. Happy Cat's ethos can be summed up by a quote on their website from author and environmental activist Wendell Berry: "We stand for what we stand on."

Mountz teamed up with Terrain Garden Café to turn his harvest into a special four-course dinner with various creative recipes incorporating the many heirloom tomatoes.

Happy Cat Farm—Kennett Square, Pennsylvania, USA

Lachlan, Tasmania, Australia

Growing an Agrarian Community

Rodney Dunn and **Séverine Demanet** exemplify the lush agrarian life through their farm, cooking school, and eatery in Tasmania's southwest.

The Agrarian Kitchen—Lachlan, Tasmania, Australia

"From the time the plane touched down I was enchanted," Rodney Dunn recalls of his first visit to Tasmania. "It was the landscape, the people, and the potential of this relatively untouched island."

Dunn, who was initially trained as a chef, was working as the food editor of Australian *Gourmet Traveller* magazine in bustling Sydney. "And I grew up on a farm, so I've always had a love of open space and greenery."

Enamored by Australia's wild southern state, he and his wife Séverine Demanet made the move to the verdant Derwent Valley in 2007. "It was Séverine's idea to create the school, which would allow us to live the life we dreamed of—growing our own produce, but more importantly, sharing that experience with others," said Dunn.

They bought five acres and transformed the on-site 1887-built schoolhouse into their family home and cooking school. The region has a rich agricultural history, and their new land was lush with fruit trees, oaks, elms, and poplars. "To establish the land for farming, we began by ploughing a large garden. The pigs dug through the soil removing the weeds, and this gave us a nice clean base in which to establish our vegetable, berry, and fruit tree plantings."

Ten years of hard work and their two cultivated acres are a haven of organic heirloom fruits, nuts, and vegetables, as well as rare breed animals. "It is full of life. Apart from the pigs, we also have geese, chickens, bees, sheep, and dairy goats. We have added two large greenhouses and a smokehouse, and converted a small open

Hands On—Rodney and Séverine host cooking and farming workshops, ranging from nose-to-tail butchery and cheese-making to paddock-to-plate foraging, harvesting, and eating.

The Agrarian Kitchen—Lachlan, Tasmania, Australia

Heritage Buildings—The Agrarian Kitchen Eatery opened in June 2017 in nearby New Norfolk's Willow Court, the town's old mental asylum.

> "We like to say we farm with common sense. We don't use anything to interrupt nature's processes."

shed into a milking shed for the goats." The berry patch contains six 30-meter (100-foot) trellises of rare berries and currants, and there are almost 70 fruit trees, from apples and peaches to cherries, lemons, and almonds. "We like to say we farm with common sense. We don't use anything to interrupt nature's processes, and we assist nature by building our soil through the use of animal and green manures. We create our own compost and all weeds are pulled by hand."

The philosophy behind their business, The Agrarian Kitchen, is to reconnect the kitchen with the land and create a place where people can rediscover the simple pleasures of gathering and cooking produce at the source of harvest. "There was a point in my life when I realized to eat better food we had to get closer to the source of that food. Distance destroys flavor. And we need to put the best food we can into our bodies to get the best outcome for our health and wellbeing." The farm-based cooking school hosts classes in nose-to-tail butchery, charcuterie, cooking with fire, cheese-making, sourdough baking, preserving, and paddock-to-plate foraging, harvesting, and eating.

In June 2017, Dunn and Demanet also started a new chapter, opening a 60-seat eatery and shop in nearby New Norfolk. Here, they celebrate local, seasonal produce, making use of the abundance of food growing on the farm and supporting the panoply of incredible growers in their community. "A true agrarian system was built upon a community, whereby a neighborhood grew and bartered with each other—one family swapping milk for another's pork or vegetables. This diversity in a neighborhood actually builds a stronger food system," Dunn says. The eatery's handcrafted wood-fired oven, grill, and smoker were built from old bricks found on site, and excess produce is pickled, jammed, fermented, or cured and then sold in the store.

"The best thing about growing a garden is that it is constantly evolving; you are never finished. There is always something to look forward to."

The Agrarian Kitchen—Lachlan, Tasmania, Australia

Human-Scale Farming Can Save the World

– Farmer and educator **Jean-Martin Fortier** is a dedicated advocate of small farms, which can sustain families and communities

Saint-Armand, Quebec, Canada

Jean-Martin Fortier believes in a simple solution to the many problems created by modern agriculture: more small-scale farmers. Farmers, educators, and authors, Fortier and his wife Maude-Hélène Desroches are the founders of Les Jardins de la Grelinette, a 10-acre microfarm in Quebec and a leading exemplar of how local human-scale farms can sustain families and communities with minimal impact on the environment. Through his work, Fortier hopes to encourage future famers to take on the fulfilling lifestyle of feeding their communities while deeply connecting to the land. A proponent of organic and biologically intensive cropping practices, Fortier aims to grow better rather than bigger, optimizing a cropping system that is both lucrative and viable. Human-scale farms are not only a great small business model but also a way for people to rediscover higher nutritional value in what they eat, which has been lost in the wake of a food industry wholly focused on quantity over quality.

Les Jardins de la Grelinette—Saint-Armand, Quebec, Canada

Through his work, Fortier hopes to encourage future famers to take on the fulfilling lifestyle of feeding their communities while deeply connecting to the land.

Fortier's farming philosophy focuses on intelligent farm design, the use of appropriate technologies, and the power of soil biology.

Les Jardins de la Grelinette—Saint-Armand, Quebec, Canada

Good Food Grown in the City

– Edgemere Farm turned an abandoned lot into a thriving organic farm

Far Rockaway, New York City, New York, USA

How did an abandoned lot for parking cars and changing tires in Far Rockaway, New York, become one of the city's most productive urban farms? A whole lot of work and dedication from a four-person team over four years.

Edgemere Farm was established in 2013 and has been growing more than 40 varieties of plants and hosting weekly farm stand markets ever since. Satisfying a demand for great produce not tainted by the woes of industrial agriculture, the farm not only delivers fresh produce to neighborhood residents, but also to Michelin-starred restaurants where purity of taste is key.

A palatial sanctum within the urban jungle, the farm also hosts dinners prepared by visiting chefs and composed of ingredients grown straight on the property for a genuine farm-to-table experience.

Organically grown vegetables, flowers, and herbs flourish on the self-sustaining half-acre urban farm on Beach 45th Street, where chickens roam freely and bees pollinate at their whim.

All profits benefit the farm directly in order to grow its projects and meet its operational costs, allowing the process of teaching and learning about healthy and delicious food to continue on its upward trajectory.

A self-sustaining and zero-profit business, Edgemere Farm also hosts volunteers on Saturdays who are encouraged to get their hands dirty and learn about growing their own food.

Satisfying a demand for great produce not tainted by the woes of industrial agriculture, the farm not only delivers fresh produce to neighborhood residents, but also to Michelin-starred restaurants where purity of taste is key.

Edgemere Farm—Far Rockaway, New York City, New York, USA

Open every weekend through October, Edgemere's farmstand is the city's most ecologically friendly market, selling the urban farm's fresh produce as well as herbs, flowers, honey, eggs, and jam.

Edgemere Farm—Far Rockaway, New York City, New York, USA

Edgemere Farm—Far Rockaway, New York City, New York, USA

Basic Pickling

Ingredients

Vegetables you have in abundance and want to preserve (onions, beets, beans, carrots, cabbage, radishes, mushrooms...)
Vinegar (white wine or apple cider)
Salt
Unrefined sugar
Spices (whatever is in your pantry: dill seeds, mustard seeds, black pepper, red chili flakes, coriander seeds...)

Equipment

2 large pots
Glass jars and lids
Kitchen tongs
Oven mitts
Towels

Instructions

1. Rinse and chop the vegetables so that they fit in the jars (and in your sandwiches).

2. In a large pot, make the pickling liquid. A good basic recipe is half water and half vinegar, some sugar, and some salt. The quantities of sugar and salt are somewhat up to taste, but you want around 3 tablespoons of sugar and 2 tablespoons of salt for every 2 cups (500 ml) of liquid. Bring it all to a boil.

3. In the meantime, sterilize the jars. Fill another large pot with water and bring it to a boil. Immerse the jars and their lids in the boiling water for about 5 minutes.

4. When your pickling liquid is boiling and your jars are ready to go, it's time to can! Using tongs, carefully remove one jar and its lid. Fill it up with vegetables and spices and top it off with the pickling liquid, leaving a finger of headspace. (Careful, everything is steaming hot! You might want to wear oven mitts.) Wipe the rim of the jar with a damp towel, and screw the lid on (not too tight). Repeat with the remaining jars.

5. Set jars to cool on a towel on your countertop. Once cool, check if the jars have sealed (the lids should not pop back up after you press them). If a jar has not sealed, place it in the fridge and use it first, or repeat the process with a new lid.

Good to Know

Pickling is a great method for preserving seasonal vegetables with (almost) all of their nutrients and flavors. The vinegar provides an acidic environment that prevents harmful bacteria from developing, and the canning method seals the whole jar to make sure it keeps in your cupboard shelves for at least one year, or until the next harvest season!

Enjoy!

The Buzz of Honey Bees

– **Michael Mutscher**'s organic honey is a dream come true

Franconian Switzerland Nature Park, Germany

Michael Mutscher's life as an organic beekeeper began with a dream. It was the sound of fervent buzzing and a deep contentment that convinced him to try his hand at honey production. Trained as a gardener, Mutscher was already attuned to the steady rhythm of the natural world.

He started with just a few bee colonies and soon had more than 30, from which he produced organic honey for direct sale. Always keeping the well-being of his bees in mind, Mutscher harvests his honey after the flowering season each year, usually in May or June. He can produce 30–40 kilograms (66–88 pounds) annually, depending on that year's weather conditions.

Wholly dedicated to organic practices, the beekeeper is fortunate to live among the pristine fields and orchards of the Franconian Switzerland Nature Park, where it is still possible to produce honey using natural, conventional practices.

The only factor that endangers the bees in Franconian Switzerland is cold weather conditions, which can freeze the orchard blossoms and make it impossible for bees to reach the nectar.

Michael Mutscher—Franconian Switzerland Nature Park, Germany

As an organic beekeeper, Mutscher must adhere to natural principles, including the use of beehive boxes made only of wood.

Michael Mutscher—Franconian Switzerland Nature Park, Germany

Honey & Nut Granola

Ingredients

4 cups (400 g) oat flakes
2/3 cup (100 g) sunflower seeds
2/3 cup (100 g) sesame seeds
2/3 cup (100 g) pumpkin seeds
2 cups (270 g) nuts of your choice (almonds, walnuts, hazelnuts, pecan nuts, cashews), roughly chopped
3/4 cup (250 g) honey
3/4 cup (150 ml) sunflower oil
Dash of cinnamon
Dash of ginger powder
Dash of salt
Handful of chopped dates or other dried fruit (optional)

Equipment

Heavy saucepan
Large mixing bowl
2 baking sheets lined with parchment paper or silicone mats

Instructions

1. Preheat oven to 170 °C (340 °F). In a large bowl, mix all the dry ingredients and set aside. In a heavy saucepan, mix the oil and honey and heat over medium heat. Remove from heat just before it boils. Pour over the dry ingredients and mix with a wooden spoon, then your hands (once cool enough) until all dry ingredients are thoroughly coated. If the mixture is too dry, add some more oil.

2. Divide the oat mixture between two parchment- or silicone-lined baking sheets and spread into an even layer. Bake for approximately 30 minutes or until golden, turning every 10 minutes with a spatula.

3. Let cool completely at room temperature, then stir in the dried fruit, if using. Store granola in an airtight container.

4. Enjoy for breakfast with some yogurt and fresh fruit or as a portable snack when you go hiking, for example. You can play around with the ingredients and use whatever you have in stock in your kitchen. Granola is also very easy to make every week with the kids. It is infinitely customizable and a great way to experiment with seasonal nuts and seeds in your area while supporting your local beekeeper. You can also use different flaked cereals and grains, depending on what grows in your region.

Good to Know

The "organic" label often doesn't mean much when it comes to honey because the bees will harvest pollen from plants within 3 kilometers (1.8 miles) around their hives no matter what. That said, buying from a local, small-scale beekeeper is the best way to get raw honey, which means it has not been heated past its pasteurization point; therefore, it retains its antioxidants and anti-fungal properties, live enzymes, vitamins, minerals, and bee pollen.

Enjoy!

A Taste of Nature's Perfume

– **Kille Enna** takes the aromas of her organic garden and translates them to the tastebuds

Ystad, Southern Sweden

Kille Enna spent seven years experimenting in her aroma atelier with raw plant materials to create her Taste of a Scent handmade aroma extracts.

It wasn't until trained chef, author, and self-taught photographer Kille Enna discovered a dusty copy of *Perfume* by Patrick Süskind in her attic that she was inspired to delve into the world of scents—more specifically, scents that can be tasted.

Having been one of the youngest head chefs in London at 21 years old under the tutelage of celebrity chef Antony Worrall Thompson, Enna moved with ease into experimentation with aromas and culinary craftsmanship.

Enna's Taste of a Scent line—handcrafted aroma extracts for drinking water—are composed purely of flowers, roots, herbs, seeds, and bark. Capturing the bold yet fleeting scent of a fully bloomed flower in her aroma atelier is only one of the challenges to which Enna has risen with avid admiration. Her latest project, an aroma cookbook, documents a fruitful yearly cycle. In Sweden's southern countryside, the profuse scents of rare herbs and medicinal plants in her cold greenhouse and organic garden provide a new opportunity with every harvest.

In Sweden's southern countryside, the profuse scents of rare herbs and medicinal plants in her cold greenhouse and organic garden provide a new opportunity with every harvest.

All of Enna's drinkable aromas are entirely organic and sustainable—from the ingredients to the glass flacons and even the cardboard used for transportation.

64   Kille Enna—Ystad, Southern Sweden

The Magic Kingdom of Seaweed

– Wild food innovator and forager **Roushanna Gray** shares her love for South Africa's coast

Cape Point, South Africa

There is food growing all around us, if only we know what to look for and where. That is the central philosophy of Veld and Sea, a workshop and event organizer that traverses land and sea to reconnect with nature's edible elements. Avidly promoting sustainable and responsible foraging, the group was founded by former urbanite Roushanna Gray, who left Cape Town to live with her husband in Cape Point, a nature reserve abounding in resplendent flora and fauna. With a view of the mountains looming grandly in the distance and surrounded by nature's full bloom, it's no wonder Gray began to feel the urge to include the landscape in her cooking.

Introducing herself to the world of indigenous plants through books and experts, she soon found herself on the shore where she began unraveling the core of her main

Gray was introduced to the wild and varied world of edible seaweed after hosting a Japanese cyclist in 2012 who had traveled the length of Africa.

focus: about 720 species of indigenous edible seaweed. Through Veld and Sea's workshops, classes, and events, elements such as the coast's enchanting rock and tidal pools, macro-algae, and shellfish become the source of an unforgettable outdoor banquet.

Respecting wildlife is a central theme for Gray, who only forages seaweed and mussel species that are already prolific and can therefore be consumed.

Veld and Sea—Cape Point, South Africa

Through Veld and Sea's workshops, classes, and events, elements such as the coast's enchanting rock and tidal pools, macro-algae, and shellfish become the source of an unforgettable outdoor banquet.

The Perfect Loaf of Bread

—Norwegian bakers **Casper André Lugg** and **Martin Ivar Hveem Fjeld** rely on nature to create their signature sourdough

Fredrikstad, Norway

One of the essential techniques Lugg and Fjeld use to make their sourdough is the stretching of the dough for about four hours at room temperature in 30–60 minute intervals.

For Casper André Lugg and Martin Ivar Hveem Fjeld, the art of making sourdough is all about taking nature's lead. With childhood memories bedecked in strong olfactory notes of freshly baked bread, their appreciation for the nourishing dietary staple led them to explore the traditional craft of sourdough-making. For the young bakers it's all about the simple combination and fermentation of flour, water, and salt combined with a knowledge and reverence for time-honored techniques. The duo integrates old grain varieties that possess more flavor and nutrition and works with farmers and small local mills that cultivate grains on sustainable terms. With patience, intuition, and a ban on kitchen machinery to aid in the process, their winning recipe values time above all.

Sourdough Starter

Ingredients

1 cup (100 g) flour (wheat or rye)
1/2 cup (130 ml) water

Equipment

Wide-mouth glass jar
Wooden spoon

Instructions

1. In a wide-mouth glass jar, mix equal parts of flour and water. Stir with a wooden spoon until well combined. The batter should be neither too runny nor too stiff, but don't worry too much about the consistency—some like their starters on the runny side, while others prefer a thicker batter. You can always adjust with a little more water or flour.

2. Cover with a cloth (or loosely cover with the lid, but it should not be tightly sealed).

3. The next day, discard half of the batter (which can be composted or fed to the chickens, if you have any) and add fresh flour and water to reach the same consistency as the day before. Let it sit again until the following day.

4. Repeat this process until your starter looks bubbly and releases aromas of beer and fruit—slightly acidic, but pleasant and not too strong. It can take between three days and a week, depending on the temperature and other factors.
If you don't notice any sign of activity after a week, discard the starter and start over from scratch.

Good to Know

A starter is a mix of flour and water that starts to ferment thanks to the live bacteria in the air. It is a living culture, and it needs to be kept warm and "fed" every day: this means adding flour and water in equal quantities, giving the starter a good stir, and letting the bacteria do their thing. To slow down the fermentation process, you can refrigerate the starter and feed it about once a week. Take it out of the fridge the day before you plan to bake, and feed it a couple of times until it becomes bubbly and active again.

A starter acts like yeast; it's a leavening agent. If you mix flour and water and bake it, it will look like a concrete brick. But if you do the same, only this time adding a little bit of starter and giving the dough time to ferment, it will rise and puff and make an airy loaf after baking! This starter can be used in any sourdough bread recipe.

Discarding and replenishing the batter speeds up the fermentation process! Once your starter is active, you won't need to discard half of it anymore, because you will be using some of it to bake with—typically around one part of starter to five parts flour—and feeding the remaining batter (the "mother," as it is called).

Sourdough is very beneficial to the body. It breaks down the toughest proteins, allowing for better digestion and absorption of nutrients.

Enjoy!

Mountain Farming at Its Best

– **Alois Mauracher** and a biodiverse pasture where cows produce the best milk

Rettenschöss, Tyrol, Austria

"Quality over quantity" reigns true at the Mannerstätter Alm, where the pasture's species diversity significantly heightens the quality of the milk, especially when it comes to the vital omega-3 fatty acid.

Every morning at 5am, when most of the town is still asleep, Alois Mauracher is already awake and milking his cows. As the sun's rays begin to alter the hues of green that cover the flowing landscape of Tyrol, the cows wander freely onto the pristine pasture, with only the surrounding mountains as their constant companion. Mauracher and his wife have been running the Mannerstätter Alm (Mannerstätter Pasture) on organic principles since 1990, and their more than 17 cows spend most of the time outside grazing freely. That's exactly why the milk tastes the way it does—the many species of herbs that flourish in the unique biodiversity of the pasture are happily consumed by the cows, giving an exceptional flavor to their milk. Filled to capacity, the heavy milk cans are carried to a cable car every morning and evening and transported to the Sennerei Hatzenstädt (Alpine Dairy), which has been making cheese since 1937.

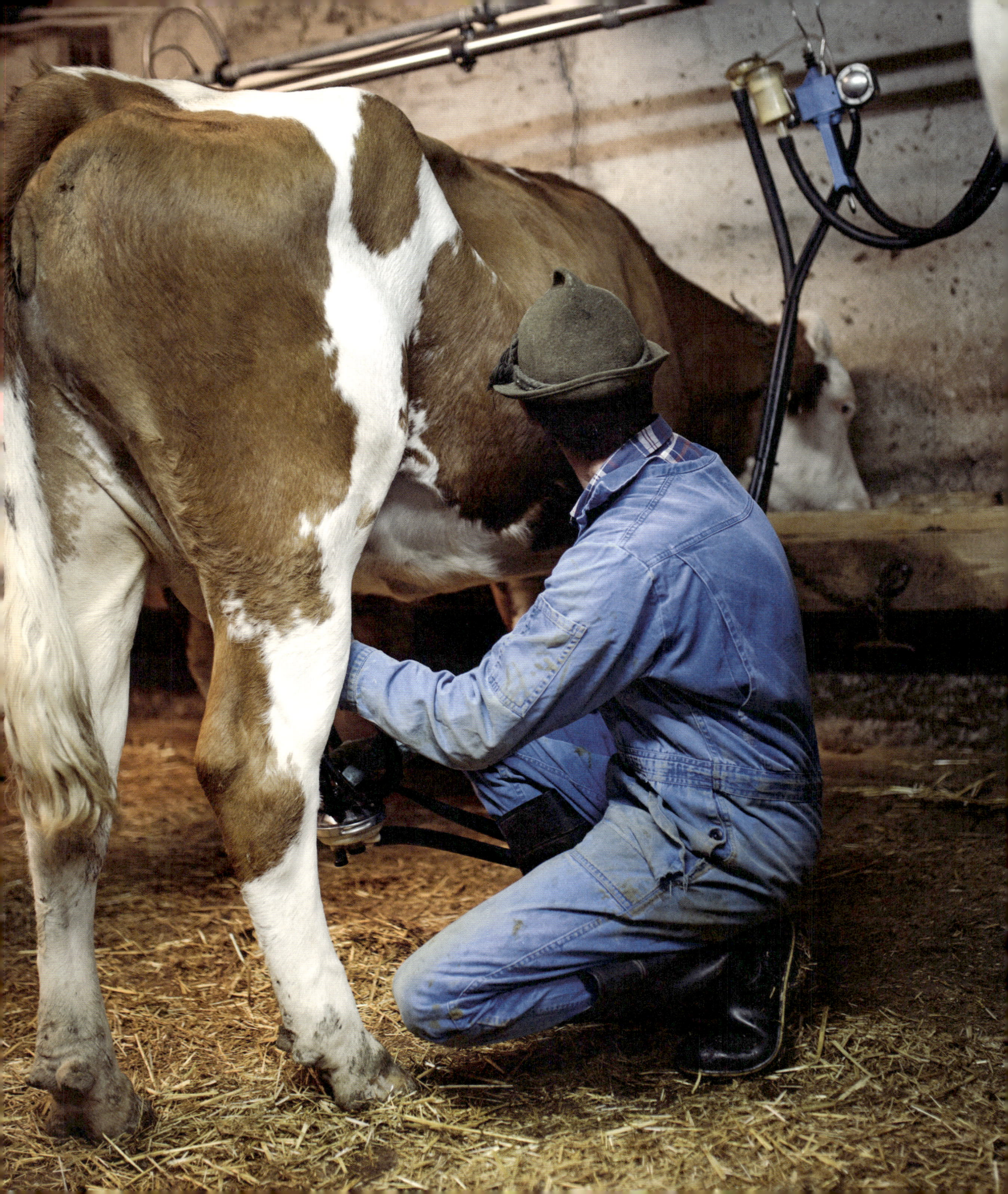

Mannerstätter Alm—Rettenschöss, Tyrol, Austria

Handmade Butter

Ingredients

Heavy cream (preferably raw, for a richer taste)
Salt (optional)

Equipment

Glass jar (for a small batch)
Stand mixer fitted with the whisk attachment (for a bigger batch)
Fine-mesh strainer
Tea towel

Instructions

1. For a small batch, fill a jar halfway with cream, then shake vigorously. Keep shaking the jar until you see knobs of yellow butter floating in liquid. This liquid is the buttermilk. Strain it, but don't discard it; it is full of nutrients, and it will happily replace water or milk when baking bread or making pancakes. For a bigger batch, whisk the cream in a stand mixer as if to make whipped cream, but push the process a little further. Butter is the next stage after whipped cream, so just keep mixing until the butter clumps separate from the buttermilk.

2. Press the butter bits together to form a ball, and place the strainer under the tap. Run cold water over the butter ball, and keep pressing and shaping to get rid of the last drops of buttermilk. It's important to rinse the butter this way, otherwise it could go rancid in just a couple of days. Rinse it until it doesn't sweat out any more buttermilk when pressed.

3. Place the butter in a clean tea towel. If using salt, sprinkle and knead it into the butter. Then, twist both ends of the towel and squeeze. Unfold, and enjoy a well-deserved tartine with handmade butter!

Good to Know

Butter is made by churning cream to separate butterfat from buttermilk. It contains a great deal of vitamins (A, D, and K2), especially when it comes from pastured cows, and around 80% fat. Yes, that's a lot of fat—but even the human brain contains at least 65% fat. No reason to be afraid of good fats!

Butter on a slice of freshly baked sourdough bread is as close to heaven as it gets. Other uses include sautéing, pan-frying, baking, melting, binding sauces, and many more.

Enjoy!

Hepburn, Victoria, Australia

Teaching a New Way of Food Consciousness

Kirsten Bradley and **Nick Ritar** live and work at Australia's original permaculture homestead, Melliodora, in Central Victoria, where they farm and teach organic agriculture.

T he name Milkwood is synonymous with organic farming and permaculture education in southeastern Australia. At the helm are Kirsten Bradley and Nick Ritar, who are well-respected stalwarts of their field. They embarked on their farming journey in 2007, and have since inspired many budding green thumbs to switch out a desk job for a small plot of land and a trowel and to integrate traditional homesteading practices into their daily lives.

Ritar has a background in software engineering, and Bradley is a musician and artist. "When we met we were making video projection art together," Bradley reflects. They were residing in inner-city Melbourne, trying to make a living as artists, when rural life beckoned.

Ritar's parents had moved to Mudgee to grow olives and graze sheep, and Bradley and Ritar decided to join them. "We thought we would build a tiny house, live simply, and make art from the farm, but then we got sideswiped by permaculture," Bradley says. "We became really passionate about it and decided to focus on hosting courses on organic farming techniques, which has spiraled into where we are today."

They chose the principles of permaculture as a framework for learning and teaching. "The design element of permaculture and designing with the whole system in mind was really attractive. Permaculture also embodied a lot of elements from our art-making practice, but farming has more concrete outcomes and is really

Pioneers—Su Dennett and David Holmgren in the garden of their 35-year-old permaculture paradise, Melliodora.

positive, so we started learning and haven't stopped." Milkwood then became what they did, rather than where they farmed. Bradley, Ritar, and their son Ashar were visiting David Holmgren, co-founder of the permaculture movement, and his partner Su Dennett at their permaculture paradise in the village of Hepburn, Central Victoria. "And a conversation started about how they wanted another family to get involved at the property. We dropped everything we were doing and moved here," Bradley says.

The property, Melliodora, is 2.5 acres of fruit trees, vegetable gardens, goats and chickens, a passive water system, and two passive solar mudbrick houses. "It is a stable ecosystem and probably the oldest example of a permaculture homestead that we have," Bradley says.

Amid this bucolic landscape, Milkwood runs on-the-ground education in permaculture design, engaging international teachers like Sandor Katz, Joel Salatin, and David Asher to lead workshops. "We're also very passionate about fostering female permaculture teachers because the gender balance is a bit wonky. Part of our ethos is to

Milkwood—Hepburn, Victoria, Australia

Above—Bradley and Ritar's son Ashar inspecting one of the farm's beehives.
Opposite—Newborn goats Rue and Mint.

Above—Ritar cultivates organic gourmet mushrooms by inoculating waste materials like straw or sawdust.

have a multitude of diverse voices teaching." The farm acts as a training ground for workshops, and also ensures that the two families are self-reliant when it comes to their personal food supply. "We focus on self-reliance rather than self-sufficiency. So we rely on ourselves rather than an outside system, but we're not striving for self-sufficiency because then we wouldn't get to see anyone else. So that inter-reliance within the community is a huge part of this place and how it operates."

Tending to the garden, spending time pickling, preserving, and cheese-making, and sharing their wealth of knowledge with the wider community are at the core of their day-to-day work. These pursuits are contributing to cultivating a higher consciousness around growing and eating that's good for the earth and for all of humanity. "I love living close to the earth and having an intrinsic relationship with the ecosystem on a daily basis. When your hands are digging in the soil, it is a visceral connection that is really grounding."

Opposite below—Homemade raw goat cheese cultured with wild kefir and wrapped in blackcurrant leaves.
Below—Son Ashar helping with the cherry harvest.

Milkwood—Hepburn, Victoria, Australia

Bulgur & Fresh Herb Salad

Recipe by Stephen Barber

Serves 4

Ingredients

1 cup (200 g) bulgur (cracked wheat)
2 sprigs fresh rosemary
2 sprigs fresh thyme
Zest and juice of **1** lemon or lime
3–4 spring onions, chopped
1/4 cup (15 g) cilantro, chopped
3/4 cup (45 g) fresh mint, chopped
1/4 cup (15 g) fresh chives, chopped
1/2 cup (30 g) fresh flat-leaf parsley, chopped
2 tablespoons (10 g) dried cherries, chopped
2 tablespoons (10 g) golden raisins
Seeds of 1 pomegranate
1/4 of an apple, diced
Some quality olive oil
Salt and freshly ground pepper to taste

Equipment

Medium saucepan
Medium mixing bowl

Instructions

1. In a medium saucepan, combine bulgur, 2 cups water, rosemary, thyme, salt, and pepper, and bring to a boil.

2. Reduce heat, cover, and let simmer for 15–20 minutes, until the bulgur is cooked and the water has been absorbed.

3. Remove the sprigs, add the lemon or lime zest and a splash of olive oil, and fluff the bulgur up with a fork. Let cool to room temperature.

4. Transfer bulgur to a medium mixing bowl, and toss with the spring onions, cilantro, parsley, mint, chives, raisins, dried cherries, pomegranate seeds, apple, lemon juice, and 2 tablespoons (30 ml) of olive oil. Taste for seasoning and serve.

Enjoy!

The Last of the Horse Fishermen

– **Stefaan Hancke** is one of the world's last horse fishermen, keeping the tradition alive

Oostduinkerke, Belgium

When Stefaan Hancke and his horse enter the sea's rolling waves early in the morning, they are actively maintaining a fishing technique that is slowly dying out.

Dragging a net, the strong legs of the calm draft horse trot into the deep water as the fisherman, clad in a bright yellow raincoat, catches scores of local grey shrimp. Horseback shrimp fishing has been a tradition for more than 500 years and Oostduinkerke is the last place where it continues to be practiced, mostly for the sake of tourism. Two days a week, Hancke sits on his horse, at one with nature, fishing for three hours before and after low tide.

Of course, the draft horse is not naturally inclined to walk right into the sea and such ability requires about a year of training, as well as a unique trust between man and animal. Despite an uncertain future, Hancke and his loyal horse continue to represent an indefatigable pride for a past more in tune with the natural world.

Stefaan Hancke—Oostduinkerke, Belgium

When fishing, horses drag a pair of boards that opens the nets in the water as well as a chain that vibrates on the seafloor and scares the shrimp into the nets. After the catch, the shrimp are pulled through a sieve in order to pick out the smallest, which are then thrown back into the sea so they can continue to grow.

First published in *The Ingredient* magazine by home appliances manufacturer NEFF.

Dragging a net, the strong legs of the calm draft horse trot into the deep water as the fisherman, clad in a bright yellow raincoat, catches scores of local grey shrimp.

Stefaan Hancke—Oostduinkerke, Belgium

Once a suitable Belgian draft horse is found and trained, the relationship between the fisherman and his horse lasts a lifetime.

Stefaan Hancke—Oostduinkerke, Belgium

From Tasmania's Pristine Waters

–From Wakame to Atlantic salmon,
James Ashmore knows seafood

Mornington, Tasmania, Australia

When James Ashmore isn't busy running his own company, he's on the open sea clad in a wetsuit hunting for wakame and native sea urchins as a commercial diver. A landscape composed of prickly pastel urchin shells and the gelatinous curves of the wakame seaweed is a common one for Ashmore, who has been working in the seafood business for years. His family-run company, Ashmore Foods, supplies some of Australia's top chefs with oysters and Atlantic salmon raised in some of the world's cleanest waters.

Ashmore also began making wakame available to the surrounding market in 2011—a seaweed native to Korea, Japan, and China, but introduced into Tasmanian waters in 1988. Retail customers can get a delicious taste of the local catch at the company's welcoming on-site market.

The Fishers of Tasmania

– **Julia and Giles Fisher** run a marine farm that is blessed with the area's pristine waters

Freycinet Peninsula, Tasmania, Australia

For Julia and Giles Fisher, the flavor and texture of fresh, locally caught seafood is the paramount prerequisite to their success. The couple has been running the Freycinet Marine Farm since 2005, as well as a small fish and chips joint that overlooks the beautiful Moulting Lagoon and the peaks of the Freycinet Mountains in the distance. While most restaurants in Australia use imported frozen fish, the Fishers only use a sustainably caught gummy shark for their crispy beer-battered variety. As for their farm, the Fishers (a rather befitting last name) grow and harvest mussels and oysters in Great Oyster Bay and sell seasonal rock lobster and fresh wild-caught Tasmanian abalone. It is the exceptionally clean waters of Tasmania's east coast that give the oysters, which travel from the leases straight onto the plate, their pure taste. The proud Fishers wouldn't have it any other way.

It is the exceptionally clean waters of Tasmania's east coast that give the oysters, which travel from the leases straight onto the plate, their pure taste.

Freycinet Marine Farm—Freycinet Peninsula, Tasmania, Australia

The Korean-Belgian Chef

– How self-taught chef **Sanghoon Degeimbre** took his extraordinary culinary career to the Belgian countryside

Liernu, Belgium

The story of Sanghoon Degeimbre is anything but ordinary. Korean-born but adopted by a Belgian family at the age of five, Degeimbre discovered a love of cooking by age 14, established a career as a successful sommelier, and went on to open his own restaurant in 1997 with no formal training as a chef. In 2012, Degeimbre and his wife moved their restaurant, L'Air du Temps, to the countryside, or, more specifically, to a beautiful farmhouse. It overlooks the fields and 12-acre garden where the central ingredients of every menu are harvested. Known for creating unique flavor combinations, the self-taught chef carefully constructs dishes that are experimental but also pay homage to the flavors of the surrounding Wallonian countryside. For an immersive experience in the restaurant and its surroundings, L'Air du Temps has five rooms that are available for weekend stays.

L'Air du Temps—Liernu, Belgium

More than 400 varieties of herbs, fruits, and vegetables are grown in the 12-acre garden, each adding their own exceptional flavor to the Michelin-starred chef's unique creations.

L'Air du Temps—Liernu, Belgium

Organic, Sustainable Luxury

– Daylesford proves that an organic, sustainable lifestyle can also be an elegant affair

Gloucestershire, U.K.

About a one-and-a-half-hour drive from London's cosmopolitan core, the English countryside appears in nature's beautifully muted tones.

It is here that Daylesford operates daily as one of the U.K.'s most organic and sustainable farms. Apart from providing their farmshops and restaurants with fresh, organic meat and poultry, produce, bread, and dairy products (from their farm, market garden, bakery, and creamery, respectfully), Daylesford has gone a step further. Illustrating that nature and luxury are not mutually exclusive, the farm's Cotswold cottages offer an organic, sustainable, and lush experience.

Pristine and modern in their decor, the converted farmhouses exude the beauty of their original features. The former wood store is now a traditional cottage with exposed stone walls and striped oak beams, while the apple store, which used to hold all of the farm's produce, is now a place where guests unwind in front of the fireplace.

Daylesford—Gloucestershire, U.K.

Apart from staying in a lovely restored cottage, guests of Daylesford Farm can also take part in an environmentally friendly floristry workshop or learn to cook seasonally and organically at the cooking school.

Illustrating that nature and luxury are not mutually exclusive, the farm's Cotswold cottages offer an organic, sustainable, and lush experience.

Daylesford—Gloucestershire, U.K.

The Tropical Farm

– Pun Pun is a pioneer in sustainable living and farming in Thailand

Chiang Mai, Thailand

About 17 kilometers (10.5 miles) off the main road, past many turns and villages steeped in the tropical flora of Thailand's unique landscape, Pun Pun makes itself known with only a small handmade wooden sign. Like a proclamation of its philosophy, the sign represents the simplicity that lies at the core of the small organic farm and learning center. Dedicated to sustainable living and the idea that learning can only be achieved through doing, Pun Pun believes that self-reliance is achieved through growing organic food, building natural homes, and saving indigenous and rare seed varieties.

Composed of adobe, bamboo, straw, and clay, the farm's houses soak up the noonday sun, while the organic garden, with its fruit trees, vegetables, and herbs, is cultivated by caring hands. Working to set an example for local farmers, Pun Pun has also opened two restaurants where traditional Thai recipes are enhanced through the use of local and nutritious ingredients.

Composed of adobe, bamboo, straw, and clay, the farm's houses soak up the noonday sun, while the organic garden, with its fruit trees, vegetables, and herbs, is cultivated by caring hands.

Pun Pun—Chiang Mai, Thailand

Above—Pun Pun's seed-saving center not only collects organic heirloom seeds but also grows, exchanges, and distributes them in order to save and propagate different species.

Left—The Pun Pun restaurants also use products from other organic farmers in Thailand, which has helped to support the farmers and grow the network.

Pun Pun—Chiang Mai, Thailand

Sweet Cherry Trees

– Orchardist **Friedrich Wölfel** continues a historic tradition with meticulous dedication

Kalchreuth, Germany

Orchardist Friedrich Wölfel is in especially high spirits when his cherries are able to survive the cold weather and mature into their dark red beauty.

There's nothing else in the world that Friedrich Wölfel would rather do than pick the ripe fruits from his orchards when the season is right. As a dedicated orchardist, he spends a lot of time meticulously cultivating and nurturing his many fruit trees, which thrive on the slopes of the Franconian Jura, 30 kilometers (18.5 miles) northeast of Nuremberg.

Wölfel grows pears and plums, but the real sensation is the dark red and perfectly sweet cherries that are famous in Franconian Switzerland, one of Europe's largest cherry cultivation areas with approximately 200,000 trees.

Like the monks of the Benedictine monastery Weißenohe, who began the longstanding tradition of fruit cultivation back in the eleventh century, Wölfel continues getting lost among the many branches heavy with the weight of mature cherries. When his basket is full, there is a myriad of various reds, each telling its own detailed account of that particular harvest season.

Seasonal Fruit Jam

Ingredients

Fruit that's in season right here, right now
Sugar, preferably unrefined
Lemon
Assorted spices (optional, see "Good to Know")

Equipment

Large mixing bowl
Wide, low pan (copper jam pans are ideal, but not mandatory)
Large pot
Glass jars and lids
Kitchen tongs
Oven mitts
Kitchen scale
Hand blender (optional)

Instructions

1. Pit the fruit, if necessary, and cut into halves or quarters. Weigh and toss fruit in a large bowl with around 60% of its weight in sugar. Mix, cover with a towel, and leave to macerate overnight at room temperature in order to release a maximum of fruit juice.

2. The next day, transfer the fruit mixture to a wide pan or a jam pan and place over low heat. Add a lemon cut in half, bring to a gentle simmer, and cook for around 40 minutes. (If using berries, you will need to skim off the scum regularly.) Take the pot off the heat, and mix jam with a fork or a hand blender, if you want a smoother texture. Be careful not to burn yourself!

3. Place the pot back on low heat while you sterilize your jars. Fill a large pot with water and bring to a boil. Immerse the jars and their lids in the boiling water for about 5 minutes. Using tongs, carefully remove one jar and its lid. Fill it up with hot jam, leaving one finger of headspace. (Careful, everything is steaming hot! You might want to wear oven mitts.) Wipe the rim of the jar with a damp towel, and screw the lid on (not too tight, just as you normally would). Repeat with the remaining jars.

4. Set the jars upside down on your countertop until completely cool. Once cooled, check if the jars have sealed (the lids should not pop back up after you press them). If a jar has not sealed, place it in the fridge and use it first, or repeat the process with a new lid.

Good to Know

When a fruit comes into season in your area, it's a good idea to buy crates full of it at the farmers' market. If you can pick the fruit yourself, even better!

Jam loves bread, bread loves jam. When out of bread, a good yogurt or hearty oat porridge can also stand in as a solid partner for the jam.

You can reduce the amount of sugar if you're not fond of using too many sweet additives, but that will also reduce the jam's shelf life. A good base ratio is six parts sugar to ten parts fruit.

Feel free to play around with spices and herbs to give your jam an extra twist. Here are a few examples of delicious combinations:
-Pear with vanilla/verbena/star anise
-Apple with cinnamon/nutmeg/rosemary
-Apricot with lavander
-Grapefruit with lemon thyme/ginger
-Strawberry with basil/mint.

Enjoy!

Happy Goats, Happy Life

– **Iain and Kate Field** left the city to establish their own Tasmanian farm

Copping, Tasmania, Australia

It was a deep discontent that drew Iain and Kate Field away from the city and onto the land. After Iain worked as a lecturer and researcher and Kate as an emergency medicine specialist, the couple changed their lives entirely by purchasing a farm with 100 cashmere goats, calling it Leap Farm. Entranced by the beauty of the Tasmanian landscape, the couple's love for the land became their livelihood, as they expanded the farm with the purchase of an additional 50 cows and one bull. By 2012, the Fields began selling their goat meat at the Bream Creek Farmers Market and by 2013, they began producing their own beef. Adhering to organic principles in order to improve the land and increase biodiversity and productivity, the couple believes the secret to quality products is happy and free-range animals. Leap Farm also has a commercial kitchen where delicious goat curry and raw goat mince with fresh herbs and spices are prepared.

After Iain worked as a lecturer and researcher and Kate as an emergency medicine specialist, the couple changed their lives entirely by purchasing a farm with 100 cashmere goats, calling it Leap Farm.

Leap Farm—Copping, Tasmania, Australia

Homemade Goat Cheese

Ingredients

1qt (1 l) goat milk
2 tablespoons apple cider vinegar
(or **4 tablespoons** lemon juice)
Salt
Assorted herbs and spices (optional)

Equipment

Medium-sized pot
Wooden spoon
Fine-mesh strainer
Cheesecloth or muslin
Mixing bowl
Cooking thermometer (optional)
Ring mold (optional)

Instructions

1. Rinse your pot briefly, and don't dry it. This will prevent milk from sticking to the bottom. Pour the milk into the pot and place on low-medium heat until it reaches 80 °C (180 °F). Lower the heat, and add the vinegar or lemon juice. Stir with a wooden spoon for another 2–3 minutes, then turn off the heat.

2. Cover the pot with a towel and let cool completely, which will take about half a day.

3. Line a strainer with cheesecloth, place it over a bowl, and pour the (now curdled) milk into it. Fold the cheesecloth over, place in the fridge or another cool place, and let sit again for about half a day.

4. To finish, fold up all four corners of the cheesecloth and gently squeeze the cheese into a ball. Season with a sprinkle of salt, pepper, or any herbs and spices that tickle your fancy. Mix them in with a fork, then reshape cheese into a ball or press into a ring mold and refrigerate overnight.

5. For a softer, spreadable goat cheese, leave the curdled milk in the cheesecloth-lined strainer for just a couple of hours or until it reaches the desired consistency. The longer it stays in the cloth, the drier the cheese will be.

Good to Know

Many small farms around the world choose to keep a few goats because they conveniently eat a wide range of plants, don't take up as much space as a cow, and give just enough milk to feed a couple of families. If you don't have your own goats, try sourcing the milk from small farms in your area or browse the organic shops and markets. Most of the time, goat milk is sold pasteurized.

Soft or hard, goat cheese adds a fresh, rich, and creamy twist to pastas, salads, soups or rolls... and the list goes on. When you start making your own goat cheese and realize just how easy it is, it won't be long until you start experimenting with adding all sorts of herbs, spices, nuts, or dried fruits.

Enjoy!

Culinary Excellence Inspired by Nature's Solitude

– Chef **Magnus Nilsson** returns to his roots to create dishes composed of nature's seasonal bounty

Jämtland County, Sweden

Solitude is the precursor to appreciation and understanding when it comes to one of the world's most exclusive dining experiences.

Chef Magnus Nilsson's restaurant, Fäviken Magasinet, stands alone in nature and relies entirely upon it to bring forth the unexpected taste culminations for which it has become famous. The restaurant is housed in a converted grain store, where simplicity allows for dishes to exude the grandeur of their tastes.

All ingredients are sourced locally through foraging, fishing, and hunting, and the menu abides by the cycles of the natural world that silently thrives outside. During

Returning home after having worked in Paris was vital for Nilsson, whose success lies in his own past, steeped in the traditions of his native Sweden.

the summer and autumn, what grows bountifully is harvested, and in the dark winter, stores are filled through pickling, salting, jellying, and bottling.

Nilsson himself grew up hunting and foraging on a 50-acre farm, and believes that creativity is intimately connected to past experience. His creation is a 20,000-acre refuge that mirrors the serenity of Jämtland's remote beauty.

Fäviken Magasinet—Jämtland County, Sweden

Chef Magnus Nilsson's restaurant, Fäviken Magasinet, stands alone in nature and relies entirely upon it to bring forth the unexpected taste culminations for which it has become famous.

Fäviken Magasinet—Jämtland County, Sweden

Getting a reservation at Fäviken Magasinet is not an easy task, but the reward is an unforgettable meal and a chance to stay overnight with breakfast included.

Fäviken Magasinet—Jämtland County, Sweden

Photographing the Purity of Nature

– Photographer and mushroom hunter
Bruno Augsburger explores the wild

Zurich, Switzerland

Total exposure to nature and the wild is what has made the photography of Bruno Augsburger so unique. The early morning mist on mountainous terrain, the dark red flash of a poisonous mushroom, or the almost still-life scene of nature at its visual peak are all captured by Augsburger's lens. The Bernese Oberland native is based in Zurich, Switzerland, and frequently ventures into Canada and Scandinavia to explore. Every spring, the photographer turns into an avid mushroom hunter, traveling to places such as Scotland, Iceland, Canada, and Alaska in search of seasonal edible fungi. Through his work, the thrill of the hunt is exuded through distinct details, and it's no surprise that Augsburger has piles of books about the fungi world to appease his vast curiosity.

Of course, the best mushroom ends up being the one that possesses the best taste, but searching for the perfect patch out in the wilderness is what makes the experience one to remember.

Every spring, the photographer turns into an avid mushroom hunter, traveling to places such as Scotland, Iceland, Canada, and Alaska in search of seasonal edible fungi.

Sperryville, Virginia, USA

A Journey toward Mindful Meat-Eating

Mike and Molly Peterson manage their herds of cows, pigs, and sheep holistically, keeping the health of the earth, the animals, and their community at heart.

Heritage Hollow Farms—Sperryville, Virginia, USA

Mount Vernon Grassfed had been in the Miller family for eight generations. Mike and Molly purchased the business from farm owner Cliff Miller III, rebranding to Heritage Hollow Farms.

"It all felt fairly natural, but even we were surprised we made the leap to being farmers."

At the base of the verdant Shenandoah National Park in Virginia is a sprawling property with pastures of diverse grasses, thriving orchards, tall forests, and wildlife corridors. It's here that Mike and Molly and their one-year-old son Alden established Heritage Hollow Farms.

The Petersons met in high school in northern Illinois and moved to Colorado together, where Mike attended culinary school and Molly studied photography. After reading Michael Pollan's *The Omnivore's Dilemma,* Mike became attuned to the importance of sourcing food locally. Driven by this newfound knowledge, the couple spent their first wedding anniversary visiting a farm to collect chicken feed. "The farmer invited us to sit on his patio and we ate fresh chèvre made from his goat milk; it felt so right, so normal, and we needed more of it," Molly recalls.

Mike was feeling burned out working as a chef. "He reached out to intern on a grass-based livestock farm thinking it would help him become a better chef by learning more about his sources," Molly says. That was in 2009, and after interning for a year, followed by a few years of managing the farm, Mike and Molly signed a five-year lease and rebranded as Heritage Hollow Farms. "It all felt fairly

Landscape—The cattle, pigs, and sheep are raised among undulating pasture of diverse warm- and cool-season grasses, orchards, forests, and meadows.

natural, but even we were surprised we made the leap to being farmers."

Eight generations of the same family had farmed this property, and the Petersons were determined to carry the torch as responsible stewards of the land, promoting diversity through ecologically sound practices and raising food with integrity. Across two leased plots encompassing 400 acres, the pair goes about the labor-intensive yet gratifying work of raising 200–350 animals.

Their grass-fed, grass-finished cattle spend their entire life on pasture, with the calves, cows, steers, and heifers all living together. "We focus on rotational grazing with the intention to allow the livestock to graze the top 20–30 percent of the plant to encourage stimulation and regrowth, but not to damage the integrity of the root structure." The sheep are a mix of Katahdin and Texel breeds. "The Katahdin sheep are great mothers and their coats don't need to be shorn because they shed naturally. The ram we have is a Texel called Brian and his genetics allow for a larger finishing size."

Then there are the pasture-raised pigs—Tamworth-Berkshire crosses, the gorgeously shaggy Mangalitsas, and Mangalitsa-Tamworth crosses. The pigs run through apple orchards and oak woods, and their food

Heritage Hollow Farms—Sperryville, Virginia, USA

Heritage—Mike's grandparents had a small dairy when he was growing up in Illinois. His grandmother still can't believe that of all her grandchildren, Mike was the one to become a farmer.

is supplemented with herbicide-free, non-GMO grain. "We don't follow one specific method or ideology. The land and the animals will tell you what they need if you're willing to observe and listen, and that comes from having a close relationship with them from day to day. Farming has an incredible number of factors and variables and there is no single perfect recipe—you have to do the best you can each month, each season, and each year."

Part of the struggle of working with nature is that it requires the couple to constantly relearn and adapt. Molly explains, "Nature is fascinating and ever-evolving, so that means we should be too. Farming requires a lot of faith and surrender."

The Petersons are passionate about changing people's perspectives on meat consumption, connecting the cut of shoulder or belly at the farm shop with the animal from which it came. It's about shaping a more sustainable food system and growing a community of informed, healthful eaters. "The core of eating is to nourish one's body, so we do our best to ensure what is being consumed is filled with as much integrity as we are able. You are what you eat."

Heritage Hollow Farms—Sperryville, Virginia, USA

Bone Broth

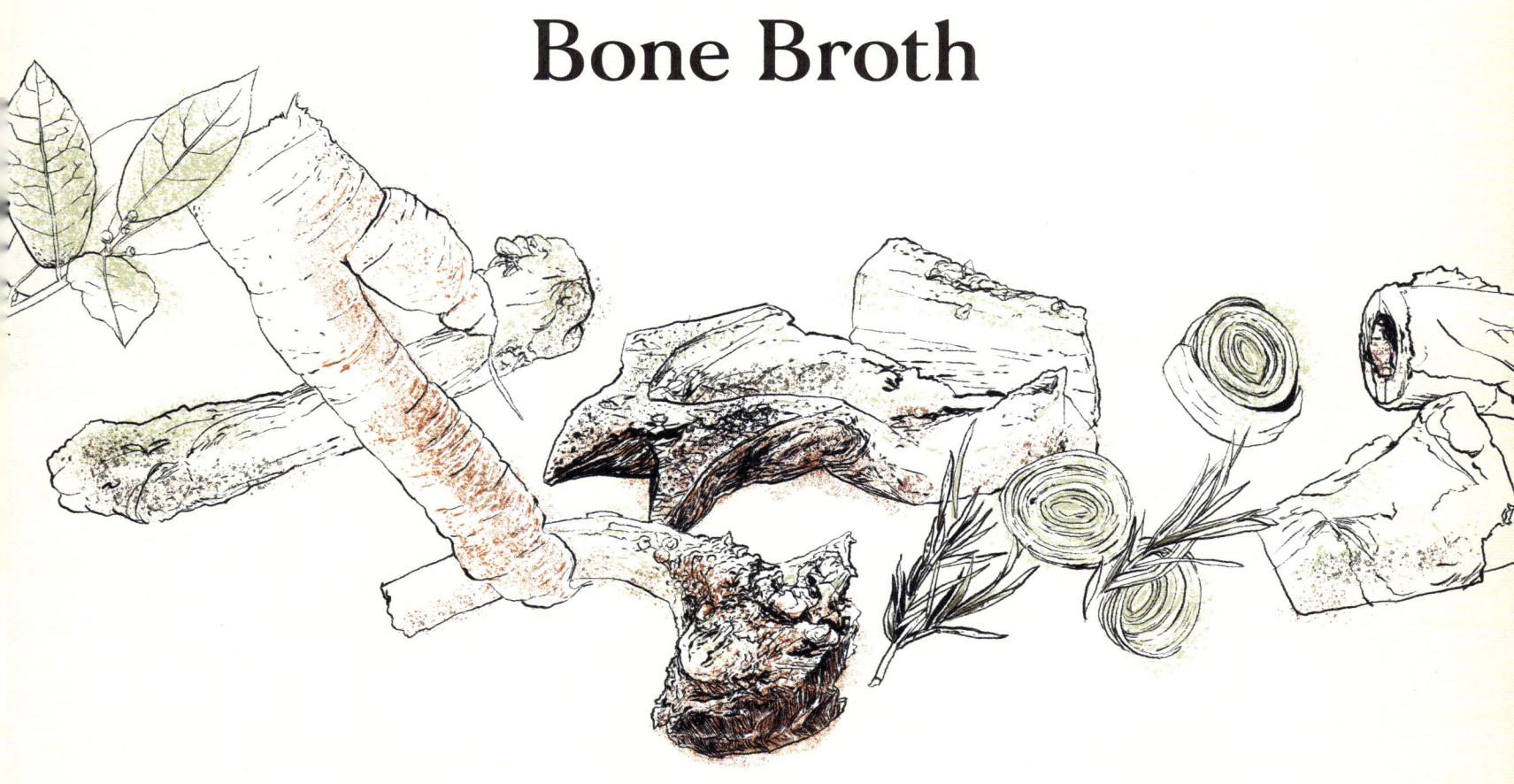

Ingredients

Bones!
(marrow, knuckles, feet, femurs...
cut up, when possible)
Anything to add flavor:
a good base is onion, garlic,
and black pepper
Sliced carrots, celery stalks, leeks, bay
leaves, and assorted spices (optional)

Equipment

Roasting pan
Large stock pot

Instructions

1. First, roast the bones (unless you are using chicken). This is an important step that dramatically improves the flavors of the broth and gives it rich layers of taste!

2. Heat oven to 225 °C (450 °F). Place bones in a roasting pan and roast until they are a deep brown color, almost burnt!

3. Then, transfer bones to a large stock pot along with whatever you have on hand for flavoring the broth. Add just enough water to cover the bones.

4. If you have a wood oven fired up, leave the stock pot on top of it for about 8 hours, or until the bones are practically falling apart. If cooking on stovetop, which is also a good way to boil down your precious liquid, bring the broth to a gentle simmer. You can simmer it for up to 24 hours, but feel free to stop whenever you think the broth is ready.

5. Nothing can really go wrong with a bone broth; the longer you cook it, the more nutrients you'll get out of those bones.

Good to Know

Bone broth is the essence of cooking. Put simply, it is bones simmered in water for a long time. A treat for the body and soul, bone broth is packed with nutrients, minerals, and gelatin, thanks to the collagen-rich joints.
It is an excellent, rich, fragrant soup on its own, but you can also make a hearty stew by boiling vegetables and/or grains in the broth. Use the broth instead of water to cook grains and cereals like rice, millet, quinoa, bulgur, or barley—you will never go back to plain water again once you've tasted the difference! Or use it as a base for sauce: sauté onions, garlic, and spices in some oil or butter until caramelized. Deglaze with a cup of broth, then blend until smooth... et voilà! A yummy sauce to go with pretty much anything savory.

Enjoy!

Floral Wonderland

– How **Fiona Haser Bizony** turned horticulture into art with the Electric Daisy Flower Farm

Bradford on Avon, U.K.

The Electric Daisy Flower Farm is like an animated impressionist painting of the most colorful sort. It comes as no surprise that its founder, Fiona Haser Bizony, is an artist and curator who gave up an office job to study horticulture. With every season, the effervescent hues and shapes of the floral world change, creating another installation carefully choreographed by the loving team at the farm.

Using only sustainable gardening practices, the harvest is not only stunning but also ethical, with a diverse group of pollinating insects and no chemicals whatsoever. At the end of the journey, every blooming flower is given new life as it becomes part of Bizony's exceptional floral designs, each telling an individual story through visual language. Composed of seasonal offerings, no two bespoke bouquets are the same, and each is deemed a small exhibition that expresses the unique elements of the chosen occasion. From Queen Anne's lace to *Verbena bonariensis*, antirrhinums to *Echinops*, Bizony's floral world is a true wonderland.

Electric Daisy Flower Farm—Bradford on Avon, U.K.

Everything at the farm is done meticulously, from the preliminary pencil sketches for every floral design to choosing the seeds, carefully selected by the flower farmers.

At the end of the journey, every blooming flower is given new life as it becomes part of Bizony's exceptional floral designs, each telling an individual story through visual language.

164   Electric Daisy Flower Farm—Bradford on Avon, U.K.

A Happier and Healthier Life

– Stowel Lake Farm is a community
and retreat where nature is ever nurturing

Salt Spring Island, British Columbia, Canada

The residents of Stowel Lake Farm have worked hard to prove that utopia is possible in an imperfect world. A tightly knit community of families that live and work together, the large estate is a perfectly organized mass of renovated eco-sustainable structures, gardens, fields, groves, and ponds, where getting lost in nature's rhythm is essential.

With space for 26 guests, the estate is also a retreat for the weary who are looking to spend time in nature, attend a spiritual and wellness workshop, or even become an apprentice on the organic farm. A profound dedication to healthy and happy lives is what drives Stowel Lake Farm, from its weekly yoga and meditation classes to the specially prepared meals composed with seasonal, fresh produce.

Whether it's the farm stand where the organically grown bounty of each season is on sale or the wall that was hand-built from the island's quarried rocks, everything on the farm exudes sincerity and love.

A significant part of the farm's ethos is its healthy cuisine prepared by its chefs, who incorporate ingredients from the gardens into their daily changing menus.

A profound dedication to healthy and happy lives is what drives Stowel Lake Farm, from its weekly yoga and meditation classes to the specially prepared meals composed with seasonal, fresh produce.

Stowel Lake Farm—Salt Spring Island, British Columbia, Canada

Chèvre-Stuffed Zucchini Blossoms

Recipe by Haidee Hart

Serves 6 as a starter

Ingredients

12 baby zucchini with blossoms attached
2 cups (240 g) fresh chèvre or ricotta cheese
3 tablespoons olive oil
Fleur de sel, for finishing

Equipment

Large non-stick sauté pan
2 small spoons
Kitchen tongs

Instructions

1. Gently open each zucchini blossom.

2. Using two small spoons, fill each blossom with chèvre or ricotta cheese. Gently twist top of blossom to close. If the blossom separates from the zucchini while you do this, just sauté it beside the zucchini.

3. Heat olive oil in a large non-stick sauté pan over medium heat. Neatly place each zucchini in the pan and sauté for about 5 minutes per side or until golden brown.

4. Using tongs, carefully remove the zucchini from the pan and arrange two on each plate. Finish with a sprinkle of fleur de sel.

Enjoy!

The Best Meat Comes from Happy Pigs

– Meine Kleine Farm is a German start-up and remedy for industrial livestock farming

Mainz, Germany

The message of Meine Kleine Farm (My Little Farm) is quite simple: eat less meat, but when you do eat it, eat it with respect. Established by Dennis Buchmann, the online butcher shop only works with small-scale, organic, and free-range farmers in order to promote a healthier relationship between consumer and product, and, ultimately, to save the world.

Every item has a sticker that features the face of the animal used to make the product as well as information about where the animal grew up and where it was slaughtered so the customer knows exactly where their food comes from. Since a high consumption of meat is not ecologically or socially sustainable, consuming less meat and doing so more sensibly is a solution that the German start-up has embraced wholly.

Happy animals, quality meat, and total transparency are what make the Leberwurst varieties and fresh cuts from free-range pigs the remedy for dubious supermarket chain products.

Quite the independent start-up with a hands-on attitude, Meine Kleine Farm produces its own pig portrait stickers for every product and is always happy to accept new sponsors.

Meine Kleine Farm—Mainz, Germany

Connect with Your Food

– Erin O'Callaghan's Rad Growers farm sends the seasonal harvest straight to its customers

Bungowannah, New South Wales, Australia

Childhood memories of feeding orphan lambs and playing in red clay channels inevitably led Erin O'Callaghan back to where she belonged. After attending university, traveling around Australia, and working in health care, O'Callaghan returned to her family's farm to immerse herself in what makes her truly happy: getting her hands back in the soil. After interning at various farms to consolidate her knowledge and experience, she established Rad Growers in 2015. Her small-scale regenerative farming project has turned into a lucrative business with veggie boxes, where seasonal subscribers receive weekly shares of chemical-free vegetables from the Rad Growers patch. Much more than a food delivery, the subscription also builds a relationship with the consumer and the farm through newsletters, recipe suggestions, farm visits, and photos.

O'Callaghan aims to share how food is produced from seed to harvest, inspiring lives that are both healthier and much more connected to the earth, where nourishment thrives.

Rad Growers—Bungowannah, New South Wales, Australia

After attending university, traveling around Australia, and working in health care, O'Callaghan returned to her family's farm to immerse herself in what makes her truly happy: getting her hands back in the soil.

Also focusing on good taste, Rad Growers researches heirloom and less common vegetable varieties to introduce new palate-based adventures to their community and customers.

Rad Growers—Bungowannah, New South Wales, Australia

Know Thy Farmer

– Clark Farm reconnects people with the land and the farmers who cultivate it with care

Carlisle, Massachusetts, USA

All the right things came together for Marjie Findlay and Geoff Freeman when they purchased this old farm on 185 Concord Street in 2010. Findlay had always been passionate about sustainable farming. The land's agricultural history dates back to the 1700s, and Freeman, a retired architect, had the right background to tackle the historic farm structures in need of restoration. Clark Farm became a project to reconnect people with the land, food, and its famers.

While actively supporting a sustainable farm ecosystem, Clark Farm also works with community-supported agriculture, an operational model where those who support the farm can pay a share upfront and receive their fresh, organic produce when it becomes available. For a more direct route to freshly grown delicacies, the market—which lies between the organic crop production fields and the seasonally available pick-your-own garden—sells the farm's produce, pasture-raised eggs, and meat. An advocate of education, Clark Farm also welcomes international students with paid apprenticeships.

Kevin Ford is one of Clark Farm's most respected names, one of the last remaining sheep shearers who prefer hand shears over the electric version.

While actively supporting a sustainable farm ecosystem, Clark Farm also works with community-supported agriculture, an operational model where those who support the farm can pay a share upfront and receive their fresh and organic produce when it becomes available.

Clark Farm—Carlisle, Massachusetts, USA

A Free-Floating Life

– Artists **Catherine King** and **Wayne Adams** spent more than 20 years building their sustainable floating home

Cypress Bay, Vancouver Island, British Columbia, Canada

In a pile of lumbered wood blown in by a storm came the sign Catherine King and Wayne Adams needed to act upon their dream of living completely unplugged and creatively immersed in nature. Freedom Cove is the name they gave their floating house, which Adams built using that very pile. More than 20 years later, the home has grown immensely. The floating complex is an amalgamation of the turquoise and fuchsia hues of King and Adams's artistic sensibilities and the deep greens of their sustainable lifestyle. It includes a dance platform off the large wooden main house, a power building with solar panels, a candle-making workshop, an art gallery, a smokehouse, a lighthouse, and a greenhouse where the couple grows their own food, embodying a philosophy based on subsistence living. For King and Adams, floating means living entirely in nature without interfering with it, and over the years they have created a deeply harmonious life with respect to their environment.

When King and Adams met in 1987 they felt an immediate soul connection heightened by their love of nature and art.

Freedom Cove—Cypress Bay, Vancouver Island, British Columbia, Canada

The floating complex is an amalgamation of the turquoise and fuchsia hues of King and Adams's artistic sensibilities and the deep greens of their sustainable lifestyle.

Freedom Cove's old fish farm system is composed of Styrofoam floats encased in hard plastic, which also became the base of more floating platforms that carry King's garden.

Freedom Cove—Cypress Bay, Vancouver Island, British Columbia, Canada

Bristol Bay, Alaska, USA

Fishing for Salmon at Graveyard Point

Fisherman **Corey Arnold** makes an annual pilgrimage to Bristol Bay, Alaska, where tens of millions of sockeye salmon arrive each year to spawn.

Corey Arnold—Bristol Bay, Alaska, USA

Community—Some of the ragtag squatters that return to Bristol Bay year after year to fish for sockeye salmon.

It's a desolate, ramshackle environment. Graveyard Point sits at the mouth of the Kvichak River, one of five that empties into Bristol Bay. Here, buildings crumble from abandonment, harsh tides erode the land, and mosquitoes are rife. In salmon season, grizzly bears come to collect any fallen catch. Around 30 million sockeye salmon are caught here annually by the 7,000 fishers that come to intercept the red fish's migration. About 130 of those men and women set up camp at Graveyard Point, bunking in run-down dormitories and dilapidated shacks at the old salmon cannery, which closed in 1959. For up to eight weeks spanning June and July, this unique community of ragtag squatters works around the clock, donning waders and lugging small skiffs of buoys, nets, and lines through muddy waters to earn their fill of wild catch. It's furious work, pulling the nets in manually, hands going numb from the repetition and the cold. It's strenuous but quiet.

Corey Arnold is one of the fishermen who returns to Graveyard Point year after year. "It feels good to be physical, off the grid, and not sitting in front of a computer. I like the challenge of pushing myself physically and mentally to make it through the season. I think we all take pride in the extreme conditions—the mud, the lack of showers, the sleepless nights, and the back-breaking heavy lifting."

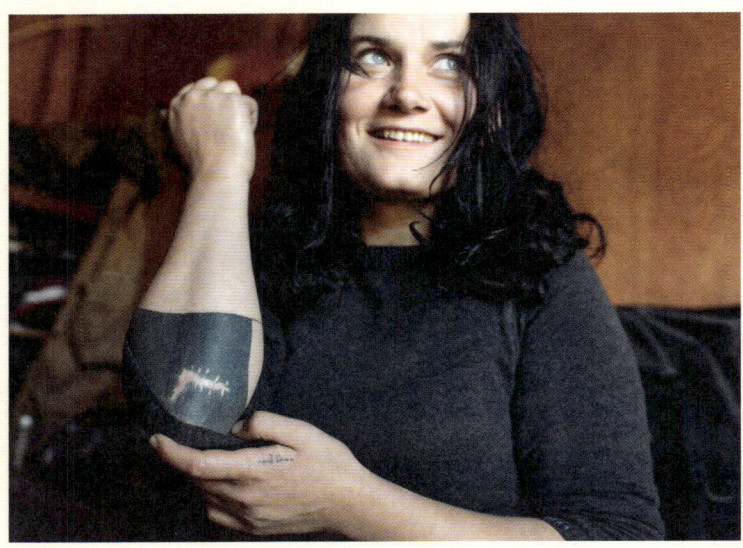

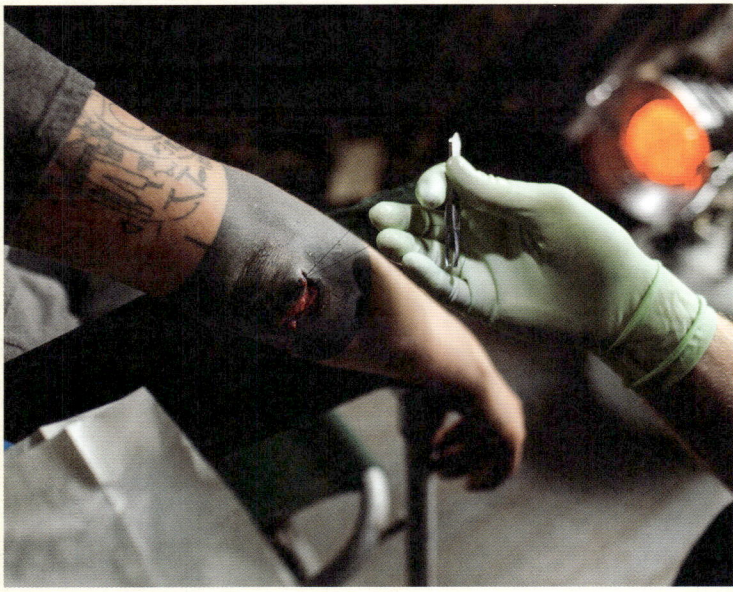

Right—Slogging through a moss-covered mudflat to explore a rumored haunted building in the distance.

Fishing is the kind of pursuit that gets in your blood. Arnold's father first took him fishing when he was two years old. Together they would spend almost every weekend angling for shark and tuna off the coast of Southern California. At 19, Arnold gained his first commercial fishing gig, a summer job in Alaska. After studying photography, he found himself back on the water, working on a Bering Sea crab boat for seven years. His documentation of this journey sparked "Fish-Work," an ongoing photography series capturing the visceral experience of life at sea.

Arnold was lured by the wild and intriguing landscape, the people, and the lifestyle found at Graveyard Point. "Community is very important. There are all types of characters to learn from that come from all walks of American life. It's a very special place with very good people and the camaraderie is strong."

He captains a commercial gill-netter, harvesting wild salmon close to shore in a delta with extreme tides. "This makes for very

Corey Arnold—Bristol Bay, Alaska, USA

> "Community is very important. There are all types of characters to learn from that come from all walks of American life. It's a very special place."

swift currents and a lot of tension is forced onto our nets, which are 300 feet long and anchored in mud on both ends," Arnold explains. Despite the ferocity offshore, he says the region is uniquely pristine, with very little human development upstream and no dams or hatcheries to supplement the supply of fish—unlike other salmon rivers in the world.

"What makes the fishery sustainable is the science-based management that keeps track of precisely how many fish are returning to spawn upstream, only allowing for the harvest of fish in excess of what is needed to sustain future runs." The fish are counted daily as they move upriver. "And though we fish with nets, we have next to zero bycatch because the delta is too silty to sustain other species," says Arnold. Most of his catch is sold to tender boats that on-sell to large processors and canneries. But a growing percentage is carefully handled and sold by his friends at Iliamna Fish Co., which is a community-supported fishery, or CSF, that sells directly to consumers in Portland, Oregon, and Brooklyn, New York. And the momentum is slowly building for the CSF model in the sprawling watershed of Bristol Bay.

Above—The harsh tide erodes the landscape and surrounding infrastructure.
Below—One of the abandoned cannery buildings has been converted into ramshackle dormitories by the fishers.

Corey Arnold—Bristol Bay, Alaska, USA

Keeping It Local on a Canadian Island

– **Leanne Lalonde** and **Jesse McCleery**'s restaurant Pilgrimme comes together through community

Galiano Island, British Columbia, Canada

On Galiano Island in Canada's British Columbia sits a wooden cabin entirely congruous with the deep green curtain of pines all around. It is here that Pilgrimme decided to establish itself as a farm-to-table restaurant that relies exclusively on ingredients from the 60-square-kilometer island.

Founder Leanne Lalonde brings experience from the hospitality industry while Jesse McCleery's comes from his culinary craft, which began at the age of 15 and culminated with a stint at Noma.

The environment is more dream home than restaurant. From the handmade ceramics used for service to the produce, grains, coffee, meats, goat's milk, and sake used for the menu, everything arrives from local producers. Whether it's using produce from a permaculture forest garden or custom-blended beans from a nearby roastery, Lalonde and McCleery work with neighbors who share their appreciation for ethical, sustainable practices.

Seasonal ingredients determine the menu at a restaurant that relies solely on its home island for inspiration.

With a deep respect for the surrounding environment, Jesse McCleery is a chef who continually experiments with the relationship between food and nature.

Pilgrimme—Galiano Island, British Columbia, Canada

A Return to Childlike Wonder

– Forager and farmer **Britt Kornum** uncovers Norway's edible weeds

Eidsvoll, Norway

Just outside of Oslo, Norway, a group of people gather at the start of each new season to wander the pastoral landscape in search of wild, edible weeds. Like carefree children, they kneel on the ground, pick weeds, and eat flowers as spring's cool wind rustles the trees.

This Get Away workshop culminates in a dinner hosted four times a year that aims to lead the local food system toward a more sustainable path. One of the foragers heading the group is Britt Kornum, an interior architect who transformed her home in Eidsvoll to make room for permaculture gardening and a few friendly goats and chickens.

Kornum knows exactly what the local landscape has to offer during each season, with flavors as wild as they are delicious, far removed from the steady palate-numbing drone of supermarket chain products.

Foraging rather than shopping, here pancakes made with fresh milk, eggs, and local spelt bubble on the frying pan and are served with homemade ricotta cheese and wild garlic pesto.

Foraging rather than shopping, here pancakes made with fresh milk, eggs, and local spelt bubble on the frying pan and are served with homemade ricotta cheese and wild garlic pesto.

It was through the New Nordic Cuisine movement, spurred on by Scandinavia's leading chefs, that an appreciation for edible plant varieties came to the fore, reintroducing foraging and challenging the status quo of food choice.

Britt Kornum—Eidsvoll, Norway

Halland, Sweden

Finding Well-Being in the Woods

Restaurateurs **Mette Helbæk** and **Flemming Hansen** know the benefits of living off the land, and they're sharing this with others from their foodie commune in a Swedish forest.

Stedsans in the Woods—Halland, Sweden

Accommodation—One of the treetop cabins designed by sustainable architects Lendager Group and made from upcycled waste materials.

Flemming Hansen and Mette Helbæk have been collaborating in the food space for 15 years—he a chef with strong farm-to-table ethics and she a food writer. Together they opened a restaurant set in a greenhouse as part of Scandinavia's first rooftop farm, ØsterGRO, in Copenhagen, Denmark. Despite the restaurant's success, they wanted to do more. "We wanted to find amazing natural surroundings so we could bring our guests out to the farm instead of bringing the vegetables to the guests," Hansen says.

It took 18 months to find the perfect forest. Set at the edge of Lake Halla in southern Sweden, Stedsans in the Woods sees over 17 acres of dense forest dotted with tiny cabins and an outdoor kitchen and restaurant. Since moving to the forest in autumn 2016, they've created a lab for a new modern lifestyle—not one that is highly

"We are animists—we believe that every living thing has a soul. This means we treat everything with respect: both animals and plants."

technology-driven, but rather that is simple and abundant in its connection with the landscape and the seasons. "The closer we are to nature, the better we feel," Hansen says. "We want to inspire people to bring more nature into their everyday life, whether that's simply growing a little food, eating clean, or using less."

Here, guests come to dine and stay. They wake immersed in the quiet of the woods, with only a sweet soundtrack of bird songs. Food

Stedsans in the Woods—Halland, Sweden

> "We created a beautiful kitchen in the middle of the forest with no electricity."

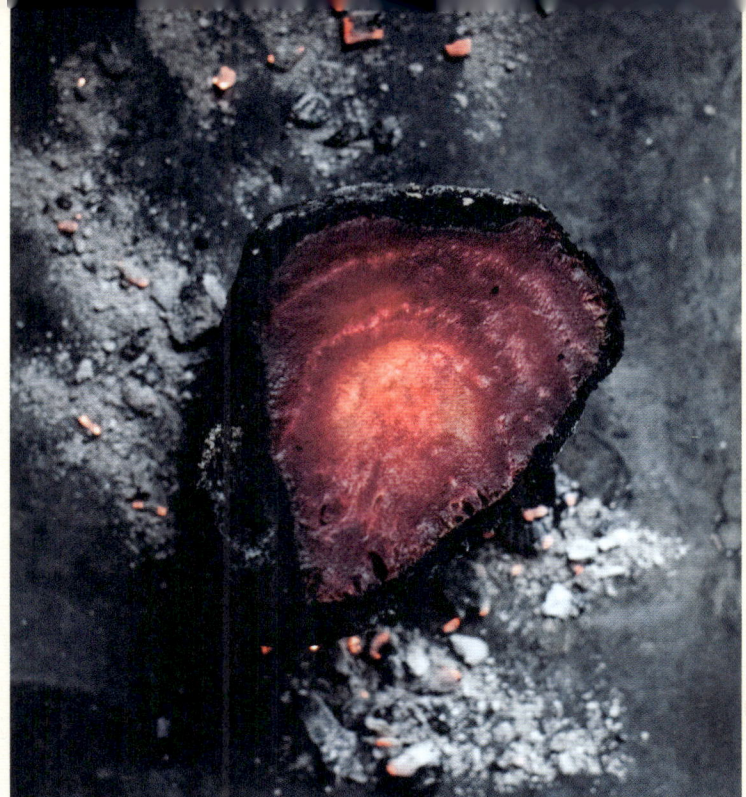

Above—A vivid beetroot cooked in smoldering coals.
Opposite—The chefs cook predominantly over fire, with ingredients harvested from the on-site permaculture garden only moments before.

is healthy and served family-style at a communal table. There is a floating sauna in a nod to the therapeutic Nordic tradition. Stedsans means "a sense of location" in Danish, which is a powerful manifesto to build on; they draw focus to the distinct terroir of their food and encourage visitors to connect with the surrounding environment and the present moment.

The centerpiece of Stedsans is the kitchen. "We created a beautiful kitchen in the middle of the forest with no electricity," Hansen says. They cook simply, often over an open fire, using vegetables from their permaculture garden and foods foraged from the forest floor—blueberries, wildflowers, edible mosses, and mushrooms. They also source from local growers. "We have so many interesting producers in our neighborhood. Within a 45-minute radius we can get fish from the lakes, crayfish from the ocean, and unpasteurized cheese. It is meaningful that we support the good people around us."

The farm is not 100 percent self-sufficient. "That would mean that we would have to skip coffee, chocolate, and wine," Hansen explains cheekily. But the team is impressively growing about 90

percent of the vegetables they need. They have 2.5 acres of vegetables and fruit, chickens, heritage-breed Linderöd pigs, and an insect hotel to prevent pests and encourage pollination. "We are animists—we believe that every living thing has a soul. This means we treat everything with respect: both animals and plants."

Ingredients are picked only hours before being cooked and served, bringing unparalleled freshness to the restaurant industry. Food scraps are fed to the animals or composted to become topsoil; gardens are watered with filtered water from the showers; the kitchen operates with minimal packaging, transportation, and fossil fuels, resulting in a low environmental footprint. "We want to create a good soil with deep roots and a lot of life. We want to live life in abundance. That means we eat good food and surround ourselves with nice things. It's about quality instead of quantity."

Hansen and Helbæk believe that sustainability and happiness are linked, which is why Stedsans in the Woods is an experience geared toward well-being for both their guests and themselves.

Above—The Third Space is the off-grid kitchen and dining area built from materials found in the forest, like upcycled wood from old barns and glass from abandoned greenhouses.

Stedsans in the Woods—Halland, Sweden

The True Farm-to-Table Experience

– How farmer and blogger **Andrea Bemis** cooks with pride

Parkdale, Oregon, USA

It can be said that everything began when Andrea Bemis moved to her boyfriend (now husband) Taylor's family farm in Massachusetts in 2008. That 60-acre organic farm was the beginning of a story filled with the joys of living simply and the rich flavors of produce grown with one's own hands. The couple went on to establish Tumbleweed Farms in Oregon, where they work the land tirelessly, harvesting ingredients for Bemis's delicious nightly dinners and acquiring a loyal following of customers. The intimate connection with nature inevitably cultivated a deep appreciation for healthy and fresh food, and Bemis began to share her recipe creations and the whole process of cooking farm-to-table on her blog *Dishing Up the Dirt*. Always working with their own seasonal and healthy produce, Tumbleweed Farms is a place where every meal contains true and unadulterated pride.

With her blog, Andrea Bemis offers an honest perspective of life on a farm, including the difficulties that are often forgotten in the fantasy.

Tumbleweed Farms—Parkdale, Oregon, USA

Farm Stand Meatballs

with Collard Greens & Apricots

Ingredients

16 oz (450 g) grass-fed ground beef
3 sprigs fresh rosemary, finely chopped
3 cloves garlic, minced
Pinch of crushed red pepper flakes
Salt and pepper
1 tablespoon olive oil
1 large bunch collard greens, stemmed and sliced into very thin ribbons (about 6 cups)
1 tablespoon grainy mustard
2 teaspoons honey
6 dried apricots, chopped

Equipment

Large bowl
Large cast-iron skillet

Instructions

1. Preheat oven to 200 °C (400 °F). Place the ground beef, rosemary, garlic, red pepper flakes, salt, and pepper in a large bowl and mix well with your hands until combined.

2. Coat a large cast-iron skillet with the olive oil. Shape the beef mixture into small meatballs, and arrange them neatly in the skillet. Bake until the meatballs are cooked through, about 18–20 minutes.

3. Using tongs, transfer the meatballs to a plate, but make sure all the pan juices remain in the skillet. Place the skillet over medium heat, add the mustard and honey, and stir with a wooden spoon.

4. Add the collard greens to the skillet and toss to coat. Cook until the greens are softened, about 3 minutes. Sprinkle in the chopped apricots, season with salt and pepper, and give the skillet a gentle shake. Finish by adding the meatballs back into the pan until they are heated through.

Enjoy!

A Small Farm with Big Bite

– Much more than a beacon for sustainable farming, Danny's Farm is a project of entrepreneurial spirit

Blampied, Victoria, Australia

For Danny Kinnear, farming epitomizes life's meaning. Connected to the land, the seasons, and his community, Kinnear's devotion to sustainable farming is both a fight against climate change and food industrialization and a privilege in making delicious and ethically produced baked goods. After traveling through Europe's organic properties and uncovering this passion, Kinnear started Danny's Farm in late 2013. What began with a herd of alpacas, Scottish highland cattle, and a garden, as well as the much-respected centrepiece—a hand-built, wood-fired pizza oven—soon expanded. Apart from raising chickens and ducks, the young farmer also keeps bees in a homemade Warre hive, forages for mushrooms, wild apples, and berries, and makes a delicious array of products, such as sourdough bread, jams, chutneys, pickles, beer, mead, and wine. Despite the farm's small size, Kinnear is able to make a living selling his food to customers at local markets, with the hope of reconnecting people with the land that sustains them.

Becoming a farmer wasn't always in the cards for Kinnear, who studied English and psychology during university.

Danny Kinnear lives as simply as possible, utilizing sustainable methods such as a reliance on rainwater for the farm's daily needs.

Connected to the land, the seasons, and his community, Kinnear's devotion to sustainable farming is both a fight against climate change and food industrialization and a privilege in making delicious and ethically baked goods.

Danny's Farm—Blampied, Victoria, Australia

Portuguese Custard Tarts

à la Blampied

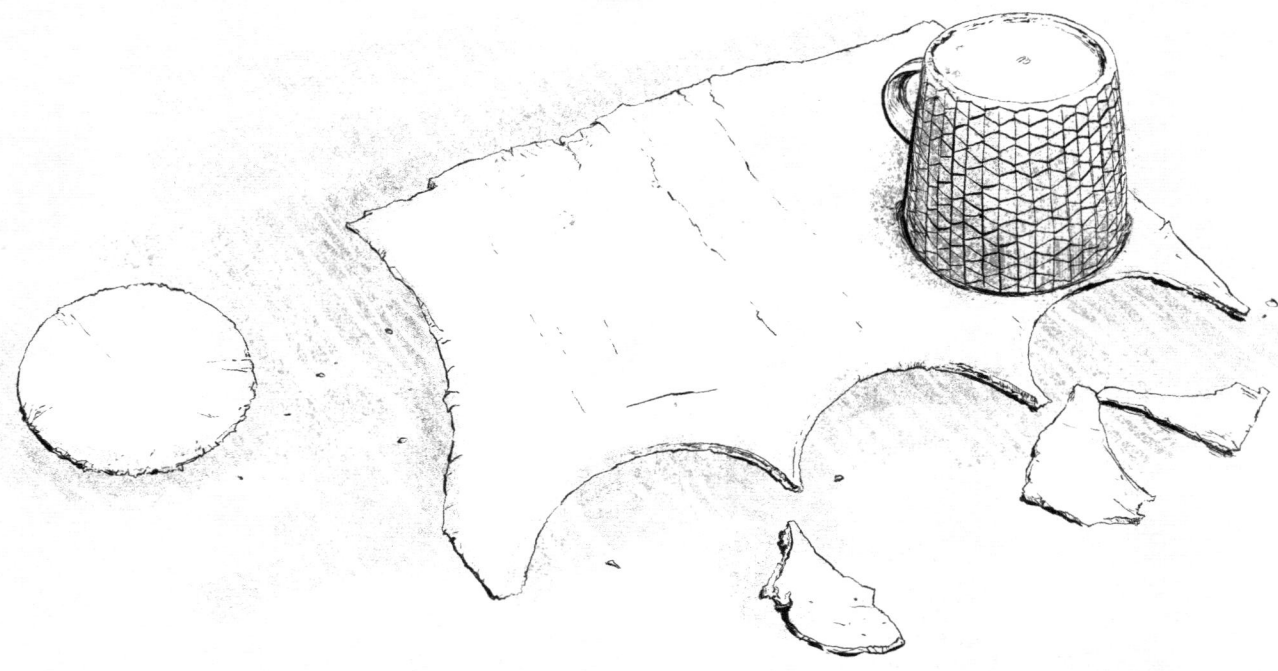

Ingredients

3 small (or 2 large) eggs
1 1/2 (350 ml) heavy cream
1 1/2 (350 ml) milk
3/4 cup (145 g) sugar
1/4 cup (25 g) corn flour
1 cinnamon stick
1 teaspoon vanilla paste
1 packet all-butter puff pastry, thawed if frozen
Butter, for greasing
Cinnamon, for dusting
Flour, for dusting

Equipment

Two 12-cup mini-tart tins
Medium saucepan
4-in (10-cm) round cookie cutter (or a cup)

Instructions

1. In a medium saucepan, crack the eggs and whisk until just combined. Slowly sprinkle the sugar and corn flour into the mix, whisking continuously. In a slow, steady stream, pour in the milk and cream, whisking well in between. Add the cinnamon stick, and place the saucepan over medium heat. Stir constantly with a wooden spoon for about 3 minutes. Turn down the heat, but keep mixing until thickened.

2. Remove the saucepan from the heat, stir in the vanilla, and let stand until cool. Preheat oven to 180 °C (350 °F) fan-forced. Grease tins with a knob of butter. Roll out the puff pastry on a lightly floured surface. Using a cookie cutter or a glass, cut out 24 4-inch rounds and place them into the cups of the mini-tart tins.

3. Discard the cinnamon stick from the cooled custard and distribute the custard evenly into the pastry shells. Bake until tops are caramelized, about 15–20 minutes. Dust with cinnamon and serve warm.

Good to Know

"Fight the temptation to eat them all at once, and remember to share the love with close friends, family, prospective employers, and Tinder dates, or swap with other handy folk for salami, cider, or home-grown veggies. Best swap so far: 20 tarts for 90 lbs (50 kg) of organic apples!" —Danny

Enjoy!

Neustrelitz, Germany

Honest Food from the Forests and Meadows

Chef **Wenzel Pankratz** plates up simple ingredients from the gardens, meadows, forests, and lakes just beyond his farm-based restaurant, Forsthaus Strelitz.

Set on 15 acres of idyllic countryside in north Germany, Forsthaus Strelitz is a farm and restaurant near the town of Neustrelitz. Growing up in his parent's restaurant spurred Wenzel Pankratz's food-centric career. In 2006 he left home to commence his apprenticeship, working in Michelin-starred restaurants in Germany, Austria, and Switzerland before returning to the family farm. "This step was always clear to me," Pankratz says.

His parents bought the farmland in 1998, where they operated a traditional inn with comforting home-style meals and accommodation. In 2014, Pankratz turned the concept on its head, applying some of his gastronomic finesse. Now guests at Forsthaus are treated to a six-course dinner with wine pairings before retiring to minimalist, design-savvy sleeping quarters in the converted barn.

Built in 1912, the barn has been transformed into eight rustic yet modern rooms, shaped by raw and natural textures. "I changed the rooms step-by-step to suit my own style. They are quite simple, honest, with carefully chosen high-quality products," says the chef. Lodgers can take a slow bath, followed by a stroll through the grounds, enjoying the benefits of nature that can be equal parts rousing and rejuvenating for a visiting city dweller.

Forsthaus Strelitz remains a family affair. Pankratz's mother takes care of the garden and the animals, his sister runs front-of-house, his girlfriend looks after bookings, and he and another chef run the kitchen. In the garden, they grow plentiful herbs, vegetables,

Below—Wenzel Pankratz makes his own pickles, preserves, vinegars, and ciders.

> "My father always tried to work as much as possible with his own produce."

and fruit, and raise chickens, sheep, ducks, pigs, geese, and bees. A polytunnel is used to mature seedlings and extend the growing season, and 98 percent of what is plated up in the kitchen comes directly from the farm. "We plant vegetables suited to this region," Pankratz explains. "When an abundance of plants is ready to harvest, we make pickles for the wintertime. We ferment, we dry mushrooms, and, furthermore, we store our carrots, beetroots, and other roots vegetables in sand, like people did back in the day. Our meat comes from our farm, except the deer and fish, which live in the forest and lakes around Forsthaus."

The chef also works with a handful of producers to source the hyperlocal ingredients not grown by his family. "A guy that we call Naked Peter is a fisher and knows a lot about mushrooms. He often brings us carp and pike." That pike is sliced into bite-sized pieces and doused in whey foam, while foraged clover, canola flowers, and Douglas fir from the surrounding woods and meadows dot each plate, capturing the tastes of the landscape in each mouthful. "My father always tried to work as much as possible with his own produce. He grew up on a farm, so he had farm life instilled in his blood. He had a vast knowledge about farming, but he passed away in January, so now there is a lot of learning by doing."

Forsthaus Strelitz—Neustrelitz, Germany

> "When an abundance of plants is ready to harvest, we make pickles for the wintertime."

Diversity in the garden translates to a year-round mélange of flavors on the plate. The small team preserves, pickles, ferments, and cures their excess harvests to ensure they are using their own produce, even in the depths of winter.

Pankratz has been touted as a leader in the Natural Kitchen movement of young chefs in Germany, cooking food in an uncomplicated manner to truly celebrate the seasonality and regional nature of the ingredients. "We don't have a particular philosophy. We just use what we have," he says. It's a return to the essentials in the kitchen, a paring back informed by traditional homesteading and historical necessity.

Forsthaus Strelitz—Neustrelitz, Germany

"We don't have a particular philosophy. We just use what we have."

Forsthaus Strelitz—Neustrelitz, Germany

Livestock Breeding with Dignity

— The pigs roam freely at Potsdamer SauenHain, a farm that fosters respect for the animals from which their high-quality products come

Potsdam, Germany

On approximately 25 acres of land, more than 200 pigs roam freely—digging, looking for food, taking naps, grunting, and sniffing around with their pink snouts. The herd's 10 mother sows and their piglets rest under the shade of the orchard's trees. Axel and Clemens gave up their 9-to-5 office jobs to intern at farms, and started this free-range pig project in 2015. Potsdamer SauenHain, which roughly translates to the Potsdam Grove of Sows, believes that the only way to acquire high-quality meat is to practice livestock

breeding with dignity. Their strictly animal-friendly husbandry means that pigs live in the great outdoors the entire year, with a diet of natural feed and whatever they can find while digging around in Potsdam's pastures and fields. The result is good meat, but also consumer respect for the sensitive and intelligent animal from which the food comes. In their online shop, Potsdamer SauenHain sells medium and large boxes filled with fresh cuts of meat and products from salciccia to ribs, bacon to cutlet, and German Wurst varieties.

Potsdamer SauenHain—Potsdam, Germany

From the Air Force to an Alpaca Ranch

– **Alvina Maynard** advocates slow fashion with her alpacas in the Kentucky hills

Richmond, Kentucky, USA

Alvina Maynard hadn't planned on becoming an alpaca rancher in the hills of Kentucky. Having grown up in a rural town east of San Diego, Maynard joined the United States Air Force right after high school, jumping out of planes and traveling the country. After leaving the military, she found herself wondering about the future, until a TV commercial for National Alpaca Farm Day led her toward a new path.

After taking some classes on alpacas and becoming thoroughly enchanted with the docile creatures, Maynard and her husband settled into Kentucky and established the River Hill Ranch. A passionate advocate of the slow fashion movement—which encourages mindful consumption and eco, ethical, and green items—Maynard produces alpaca meat as well as clothing and accessories from alpaca wool. With her two kids in tow, both eager helpers, Maynard often sits on her front porch, overlooking the farm and feeling quite blessed.

A passionate advocate of the slow fashion movement—which encourages mindful consumption and eco, ethical, and green items—Maynard produces both alpaca meat as well as clothing and accessories made from alpaca wool.

Alvina Maynard is a proud member of Homegrown by Heroes, a label for agricultural products produced by U.S. military veterans, which was started by the Farmer Veteran Coalition.

River Hill Ranch—Richmond, Kentucky, USA

Index

FOOD STUDIO
Norway
www.foodstudio.no
pp. 2–9
PHOTOGRAPHY: Sebastian Dahl

TAVSTA FARM/
LINNÉA AND PELLE HOLST
Sweden
pp. 10–21
PHOTOGRAPHY: Matilda Hildingsson,
Babes in Boyland
STYLIST: Nathalie Myrberg,
Babes in Boyland

SPANNOCCHIA
Italy
www.spannocchia.org
pp. 22–27
PHOTOGRAPHY: Valery Rizzo

HAPPY CAT FARM/
TIM MOUNTZ
USA
store.happycatorganics.com
pp. 28–31
PHOTOGRAPHY: Valery Rizzo

THE AGRARIAN KITCHEN/
RODNEY DUNN AND
SÉVERINE DEMANET
Australia
www.theagrariankitchen.com
pp. 32–39
PHOTOGRAPHY: Adam Gibson

pp. 40–41
PHOTOGRAPHY: Svein Gunnar Kjøde

LES JARDINS DE LA
GRELINETTE
JEAN-MARTIN FORTIER
Canada
www.themarketgardener.com
pp. 42–45
PHOTOGRAPHY: Alexandre Chabot

EDGEMERE FARM
USA
www.edgemerefarm.org
pp. 46–53
PHOTOGRAPHY: Valery Rizzo

PICKLES
p. 54
PHOTOGRAPHY: Kathrin Koschitzki

MICHAEL MUTSCHER
Germany
pp. 56–59
PHOTOGRAPHY: Kathrin Koschitzki

HONEY GRANOLA
p. 60
PHOTOGRAPHY: Samantha Linsell
www.samanthalinsell.com

KILLE ENNA
Sweden
www.killeenna.com
pp. 62–65
PHOTOGRAPHY: Kille Enna,
portraits: Carsten Seidel.
With thanks to Prestel Publishing.

VELD AND SEA/
ROUSHANNA GRAY
South Africa
www.veldandsea.com
pp. 66–71
PHOTOGRAPHY: Sacha Specker—66,
69 (bottom), 70 (bottom left);
Aiden Delport—67, 70 (top, bottom
right), 71, 68, 69 (top)

ILLE BRØD/
CASPER ANDRÉ LUGG AND
MARTIN IVAR HVEEM FJELD
Norway
www.instagram.com/illebrod
pp. 72–75
PHOTOGRAPHY: Simen Grytøyr
Courtesy of Modern Books,
Sourdough, 2017

MANNERSTÄTTER ALM/
ALOIS MAURACHER
Austria
pp. 76–79
PHOTOGRAPHY: Kathrin Koschitzki

TASMANIAN BUTTER
p. 80
PHOTOGRAPHY: Lean Timms
Tasmanian Butter Co: www.tasmanian-butterco.com.au

MILKWOOD/
KIRSTEN BRADLEY
AND NICK RITAR
Australia
www.milkwood.net
pp. 84–95
PHOTOGRAPHY: Milkwood—84, 86–95,
Phu Tang—85

BULGUR SALAD
Long Meadow Ranch,
St. Helena, California, USA
pp. 96–97
PHOTOGRAPHY: Long Meadow Ranch
Recipe by Chef Stephen Barber at
Farmstead at Long Meadow Ranch

STEFAAN HANCKE
Belgium
pp. 98–103
PHOTOGRAPHY: Katharina Bohm
(KatharinaBohm.de), commissioned
by NEFF's *The Ingredient* magazine.
Go to www.neff-home.com/
for more information.

ASHMORE FOODS/
JAMES ASHMORE
Australia
www.ashmorefoods.com.au
pp. 104–105
PHOTOGRAPHY: Lean Timms

FREYCINET MARINE FARM/
JULIA AND GILES FISHER
Australia
www.freycinetmarinefarm.com
pp. 106–111
PHOTOGRAPHY: Lean Timms

L'AIR DU TEMPS/
SANGHOON DEGEIMBRE
Belgium
www.airdutemps.be
pp. 112–115
PHOTOGRAPHY: Pieter d'Hoop

DAYLESFORD
U.K.
www.daylesford.com
pp. 116–119
PHOTOGRAPHY: Daylesford Organic

PUN PUN
Thailand
www.punpunthailand.org
pp. 120–125
PHOTOGRAPHY: James MacDonald
Photography

FRIEDRICH WÖLFEL
Germany
pp. 126–127
PHOTOGRAPHY: Kathrin Koschitzki

JAM
p. 128
PHOTOGRAPHY:
zoryanchik/stock.adobe.com

LEAP FARM/
IAIN AND KATE FIELD
Australia
www.leapfarm.com.au
pp. 130–135
PHOTOGRAPHY: Lean Timms

GOAT CHEESE
p. 136
PHOTOGRAPHY:
Sunny Forest/stock.adobe.com

FÄVIKEN MAGASINET/
MAGNUS NILSSON
Sweden
www.favikenmagasinet.se
pp. 138–143
PHOTOGRAPHY: Courtesy of Fäviken Magasinet—138–139, 143; stills from a film by Jacopo Cinti for Mr Porter; Director of Photography: Erik Nordlund; Producer: Marie Belmoh—140–142

BRUNO AUGSBURGER
Switzerland
www.brunoaugsburger.com
pp. 144–149
PHOTOGRAPHY: Bruno Augsburger

HERITAGE HOLLOW FARMS/
MIKE AND MOLLY PETERSON
USA
www.heritagehollowfarms.net
pp. 82–83, 150–159
PHOTOGRAPHY: Lise Metzger—150; Mike and Molly Peterson—82–83, 151–159

BONE BROTH
p. 160
PHOTOGRAPHY: Nina Sahraoui

ELECTRIC DAISY
FLOWER FARM/
FIONA HASER BIZONY
U.K.
www.electricdaisyflowerfarm.co.uk
pp. 162–165
PHOTOGRAPHY: Alma Haser

STOWEL LAKE FARM
Canada
www.stowellakefarm.com
pp. 166–171
PHOTOGRAPHY: Syd Woodward

MEINE KLEINE FARM/
DENNIS BUCHMANN
Germany
www.meinekleinefarm.org
pp. 172–175
PHOTOGRAPHY: Marcus Werner

RAD GROWERS/
ERIN O'CALLAGHAN
Australia
www.radgrowers.com
pp. 176–183
PHOTOGRAPHY: Georgie James

CLARK FARM/
MARJIE FINDLAY AND GEOFF FREEMAN
USA
www.clarkfarmcarlisle.com
pp. 184–187
PHOTOGRAPHY: Mollie McPhee

FREEDOM COVE/
CATHERINE KING AND
WAYNE ADAMS
Canada
pp. 188–193
PHOTOGRAPHY: Taehoon Kim

COREY ARNOLD
USA
www.coreyfishes.com
pp. 194–203
PHOTOGRAPHY: Corey Arnold

PILGRIMME/
LEANNE LALONDE AND
JESSE MCCLEERY
Canada
www.pilgrimme.ca
pp. 204–207
PHOTOGRAPHY: Grant Harder

BRITT KORNUM
Norway
pp. 208–211
PHOTOGRAPHY: Svein Kjøde

STEDSANS/
METTE HELBÆK AND
FLEMMING HANSEN
Sweden
www.stedsans.org
pp. 212–225
PHOTOGRAPHY: Preston Drake-Hillyard—212–213; Lendager Group—216; Stine Christiansen—214–215, 217–225

TUMBLEWEED FARM/
ANDREA BEMIS
USA
www.dishingupthedirt.com
pp. 226–231
PHOTOGRAPHY: Kate Schwager

DANNY'S FARM/
DANNY KINNEAR
Australia
www.dannysfarm.com.au
pp. 232–237
PHOTOGRAPHY: Juanita Broderick

FORSTHAUS STRELITZ/
WENZEL PANKRATZ
Germany
www.forsthaus-strelitz.de
pp. 238–247
PHOTOGRAPHY: Daniel Cramer
www.danielcramer.com

POTSDAMER SAUENHAIN
Germany
www.potsdamer-sauenhain.de
pp. 248–249
PHOTOGRAPHY: Potsdamer Sauenhain—248 (top and bottom); Vom Einfachen das Gute—248 (middle), 249

RIVER HILL RANCH
ALVINA MAYNARD
USA
www.riverhillranch.us
pp. 250–253
PHOTOGRAPHY: Meg Wilson

LE MONDE
DES MILLE COULEURS
Belgium
endpapers
PHOTOGRAPHY: Siska Vandecasteele (front); Heikki Verdurme (back)

Thanks

A big part of our collective has contributed to this book; Nina Sahraoui, the French knowmad, followed our adventures all over the world with recipes, how-tos, and by connecting to farmers she visited on her journey to learn from the pioneers of the permaculture movement in Australia; the photographers Svein Kjøde and Sebastian Dahl provided beautiful images; Caroline Hargreaves, living systems diplomat and sustainability adventurer based in Nepal shared global connections.
Bent Schrader, who helped writing the introduction, and Mira Beckstrøm Laurantzon, as parts of the introduction are based on her texts. Other valuable contributors include Rita Platiel, Pauline Lemberger, Troels Rosenkrantz, Guillermo Vázquez Bustelo, Rita Paltiel, Lise Kvan, Runi Ellingsgaard, and Megan Guertner.

Cecilie Dawes, Food Studio

Farmlife

From Farm to Table and New Country Culture

This book was conceived, edited, and designed by Gestalten.

Edited by Robert Klanten and Caroline Kurze
Contributing Editor: Food Studio

Preface by Cecilie Dawes and Rachel Sampson
Project texts by Feride Yalav-Heckeroth
Feature texts by Linsey Rendell
Recipes (pp. 55, 61, 75, 81, 129, 137, 161) by Nina Sahraoui

Illustrations by Florian Bayer

Project Management by Sonja Altmeppen and Sina Kernstock

Creative Direction of Design by Ludwig Wendt
Layout by Anna Berge, Stefan Morgner and Mona Osterkamp
Cover Design by Ludwig Wendt

Typeface: Morion by David Einwaller

Cover photography by Alissa Hessler, urbanexodus.com
Alissa and Jacob Hessler posing for an American Gothic imitation for their wedding invitation.

Printed by Printer Trento S.r.l., Trento, Italy
Made in Europe

Published by Gestalten, Berlin 2018
ISBN 978-3-89955-918-7

© Die Gestalten Verlag GmbH & Co. KG, Berlin 2018
All rights reserved. No part of this publication may be reproduced or transmitted in any form or by any means, electronic or mechanical, including photocopy or any storage and retrieval system, without permission in writing from the publisher.

Respect copyrights, encourage creativity!

For more information, and to order books, please visit www.gestalten.com.

Bibliographic information published by the Deutsche Nationalbibliothek. The Deutsche Nationalbibliothek lists this publication in the Deutsche Nationalbibliografie; detailed bibliographic data are available online at http://dnb.d-nb.de.

None of the content in this book was published in exchange for payment by commercial parties or designers; Gestalten selected all included work based solely on its artistic merit.

This book was printed on paper certified according to the standards of the FSC®.